THE POETRY OF SCIENCE:

THE POETRY FRIDAY ANTHOLOGY® FOR SCIENCE FOR KIDS

OTHER POETRY ANTHOLOGIES COMPILED BY VARDELL & WONG

The Poetry Friday Anthology for Celebrations
(Teacher/Librarian Edition and Children's Edition)
Holiday Poems for the Whole Year
in English & Spanish

The Poetry Friday Anthology
(K-5 Edition for Teachers and Parents)
Poems for the School Year
with Connections to the Common Core
or
with Connections to the TEKS

The Poetry Friday Anthology for Science
(K-5 Edition for Teachers and Parents)
Poems for the School Year
Integrating Science, Reading, and Language Arts
with Connections to the Next Generation Science Standards

The Poetry Friday Anthology for Middle School
(Teacher Edition for Grades 6-8)
Poems for the School Year
with Connections to the Common Core
or
with Connections to the TEKS

POMELOBOOKS.COM

THE POETRY OF SCIENCE:

THE POETRY FRIDAY ANTHOLOGY® FOR SCIENCE FOR KIDS

COMPILED BY

SYLVIA VARDELL & JANET WONG

WITH ILLUSTRATIONS BY FRANK RAMSPOTT & BUG WANG

pomelo * books

A NOTE TO PARENTS AND TEACHERS

This book can accompany the Teacher's Edition of *The Poetry Friday Anthology for Science*, a collection of 218 poems by 78 different poets for children in grades Kindergarten through fifth grade. **In that book, you'll find *Take 5!* mini-lessons for presenting each and every poem.**

Here, **we present the poems for all the grade levels together (K-5), but without the mini-lessons,** so that children can simply enjoy the poems. In addition, you'll find thirty "bonus" poems here that did not appear in *The Poetry Friday Anthology for Science* (K-5 Teacher's Edition). Some poems even have Spanish translations provided by the poet. For more information about this book and *The Poetry Friday Anthology* series, go to PomeloBooks.com.

WE DEDICATE THIS BOOK WITH GRATITUDE
TO THE 78 POETS
WHO
GENEROUSLY SHARED THEIR WORK
AND TO ARTISTS FRANK RAMSPOTT & BUG WANG

Pomelo Books
4580 Province Line Road
Princeton, NJ 08540
PomeloBooks.com
info@PomeloBooks.com

Illustrations by Frank Ramspott except for the following by Bug Wang: pages 13 (bird), 20 (gloves), 26 (cat), 33 (celery), 36 (heart), 39 (hermit crab), 42 (pumpkins), 49 (truck), 56 (wagon), 65 (conch), 86 (storm), 88 (landslide), 97 (trilobite), 98 (compost), 109 (bee), 111 (alligator), 113 (frog), 115 (cicada), 116 (sloth, snake), 121 (tortoise), 122 (manatee), 124 (feet), 126 (peanut), 135 (driftwood), 148 (meter stick), 155 (space yacht, sea shuttle).

Library of Congress Cataloging-in-Publication Data is available.

ISBN 978-1-937057-98-5

Please visit us:
PomeloBooks.com

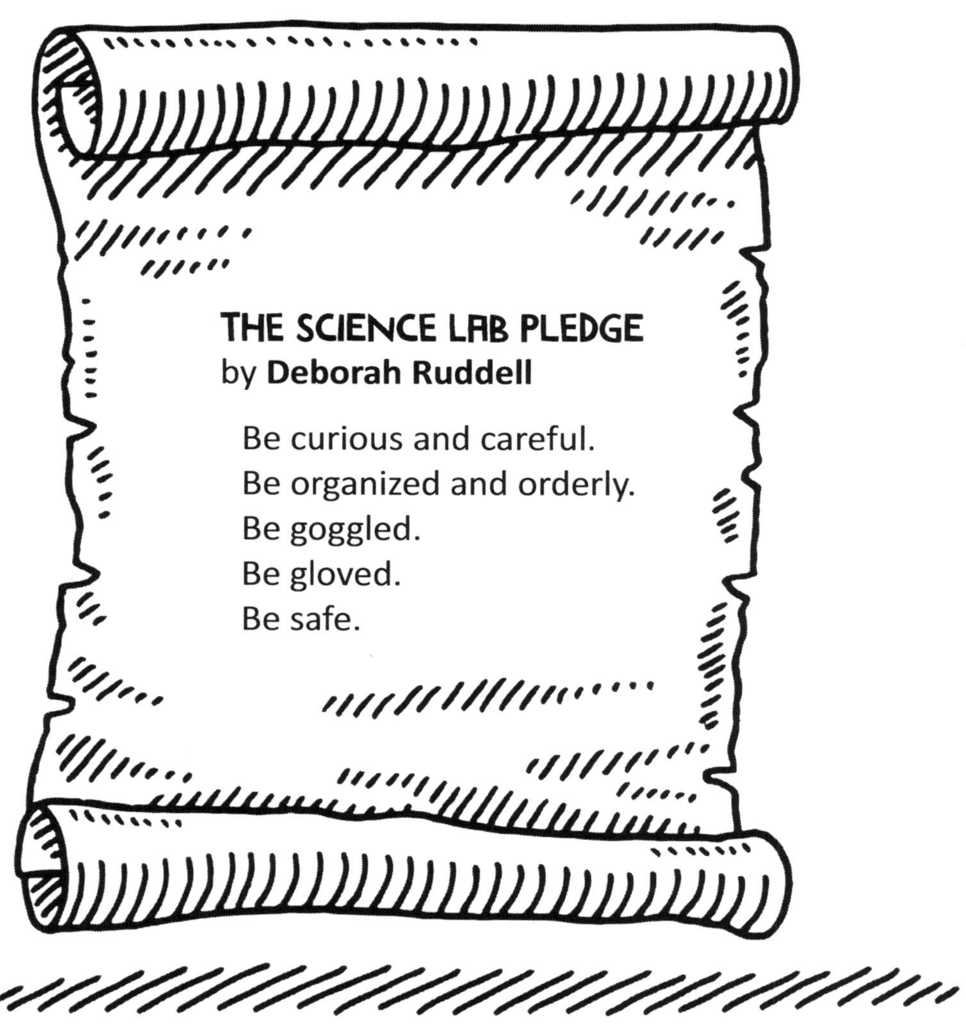

THE SCIENCE LAB PLEDGE

by **Deborah Ruddell**

Be curious and careful.
Be organized and orderly.
Be goggled.
Be gloved.
Be safe.

Joy Acey
Alma Flor Ada
Linda Ashman
Jeannine Atkins
Carmen T. Bernier-Grand
Robyn Hood Black
Susan Blackaby
Susan Taylor Brown
Joseph Bruchac
Leslie Bulion
Stephanie Calmenson
F. Isabel Campoy
James Carter
Kate Coombs
Cynthia Cotten
Kristy Dempsey
Graham Denton
Rebecca Kai Dotlich
Shirley Smith Duke
Margarita Engle
Douglas Florian
Betsy Franco
Carole Gerber
Charles Ghigna
Joan Bransfield Graham
Mary Lee Hahn
Avis Harley
David L. Harrison
Terry Webb Harshman
Juanita Havill
Esther Hershenhorn
Mary Ann Hoberman
Sara Holbrook
Patricia Hubbell
Jacqueline Jules
Bobbi Katz
X. J. Kennedy
Julie Larios
Irene Latham
Renée M. LaTulippe
Debbie Levy
J. Patrick Lewis
George Ella Lyon
Guadalupe Garcia McCall
Heidi Mordhorst
Marilyn Nelson
Kenn Nesbitt
Lesléa Newman
Eric Ode
Linda Sue Park
Ann Whitford Paul
Greg Pincus
Mary Quattlebaum
Heidi Bee Roemer
Michael J. Rosen
Deborah Ruddell
Laura Purdie Salas
Michael Salinger
Glenn Schroeder
Joyce Sidman
Buffy Silverman
Marilyn Singer
Ken Slesarik
Eileen Spinelli
Anastasia Suen
Susan Marie Swanson
Carmen Tafolla
Holly Thompson
Amy Ludwig VanDerwater
Lee Wardlaw
Charles Waters
April Halprin Wayland
Carole Boston Weatherford
Steven Withrow
Allan Wolf
Virginia Euwer Wolff
Janet Wong
Jane Yolen

NOW . . .

by **James Carter**

The birth of a star.
The beat of a heart.

The arc of an hour.
The bee and the flower.

The flight of a swan.
The weight of the sun.

A river in flood.
The nature of blood.

The future in space
for this human race.

Now that's
what I call
s c i e n c e

WHEN YOU ARE A SCIENTIST
by **Eric Ode**

When you are
a scientist,
ask what
and when
and how
and where
and why, why, why.

When you are
a scientist,
read,
and watch,
and think,
and write,
and try, try, try.

HOW TO BE A SCIENTIST
by **Amy Ludwig VanDerwater**

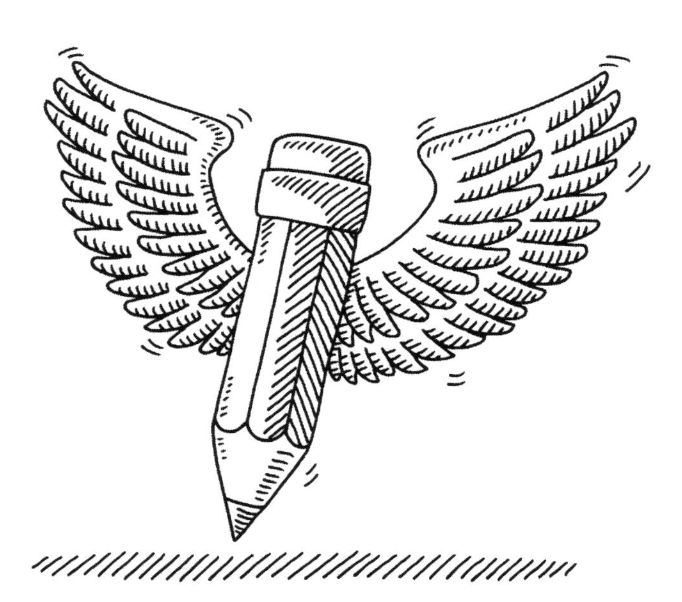

Wonder.
Ask.
Hypothesize.
Experiment.
Open your eyes.
Watch.
Now write.
What do you see?
Share
everything
you learn
with me.

SCIENTIFIC STEPS

by **Cynthia Cotten**

Find a problem, ask a question.
That's the way you start.
Now do research, all you can.
That's the second part.
Predict an answer—a hypothesis—
Based on what you know.
Run a test—an experiment.
Finished? How'd it go?
Repeat that step a few more times.
Are your results the same?
Analyze your data—
Does it back up your claim?
Write up your observations.
Be clear, avoid confusion.
Then share with others what you've found.
This is your conclusion.

SEEKING PROOF

by **Carole Gerber**

What is it I *think* I see?
What reasons make good sense to me?
Can I show my thoughts are true?
If not, I'll talk them out with you.
I want to hear what you think, too!
We'll work together to find proof
that our ideas are the truth.

DISCOVERY/DESCUBRIMIENTO

by/por **Margarita Engle**

Silence.
Careful.

We can't be noisy
if we want to see
shy birds.

I never knew
that explorers
have to be
so patient.

Look!
Such a beautiful
hummingbird!

The wait
is worth it!

Silencio.
Cuidado.

No podemos hacer ruido
si queremos ver
pájaros tímidos.

Yo no sabía
que los exploradores
tienen que ser
tan pacientes.

¡Mira!
¡Que colibrí
tan lindo!

¡Vale la pena
esperar!

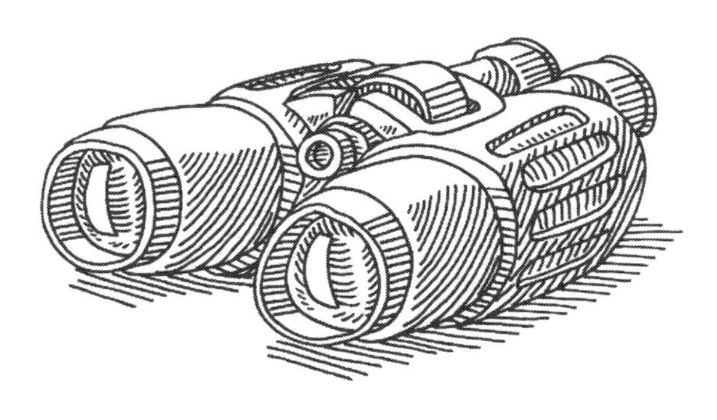

STEP OUTSIDE. WHAT DO YOU SEE?

by **Allan Wolf**

Step outside. What do you see?
A bird. A bug. A bumblebee.
A leaf. A breeze. A budding tree.

Step outside. The weather's fine.
A rock. A stick. A tall green pine.
Let's go out, rain or shine!

RACHEL CARSON

by **Julie Larios**

Loving the earth—
its deep sea water,
its wide blue skies—
she listened for the sound of birds,
hoping she could help them
keep singing, hoping
we would never have
a silent spring.

LET'S ALL BE SCIENTISTS!

by **Renée M. LaTulippe**

Let's choose a subject that we love.
Let's ask a thoughtful question.
Let's guess at what the answer is.
Let's plan the investigation.

Let's make a great experiment.
Let's watch and write and learn.
Let's see—did we find answers? Yes!
Let's show what we know to the world!

JANE GOODALL BEGINS A SPEECH

by **Susan Marie Swanson**

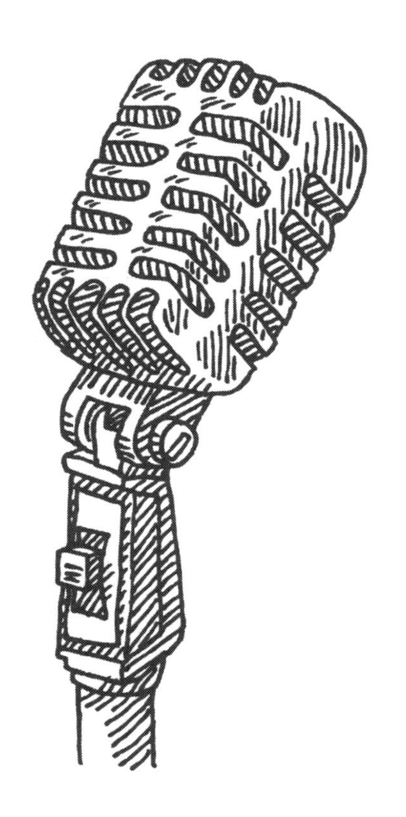

Picture a grand auditorium
lit with glittering chandeliers.
A scientist steps up to the microphone.
"Good morning," she says. "Listen."
The place is hushed.

Then, she starts to hoot and whoop,
loud and louder, shrieking,
turning the whole room wild!
From a forest in Tanzania
she has brought a greeting
from the wild chimpanzees!

The crowd claps and claps
for Jane Goodall
and her chimpanzee hello.
They clap for the stories she will tell
about the chimps at Gombe
and how she became a field biologist
who carries chimpanzee voices
all over the world.

RINGS NOT LETTERS
by **Juanita Havill**

A tree writes the story of its life
in rings not letters.
One tiny ring at the center:
"Here is where I began."
Next year a new ring:
"Look how much I grew."
Wide bands between rings:
"Hooray for rain and sun."
Narrow bands:
"It's hot and dry and I'm so thirsty."
Fires, insects, the weight
of a fallen tree against the trunk,
all written in rings, not letters,
the life story of a tree.

Note: This poem was inspired by information from the Tree Ring Lab at the University of Arizona (www.ltrr.arizona.edu) and an image and explanation at ArborDay.org/trees/ringsLivingForest.cfm.

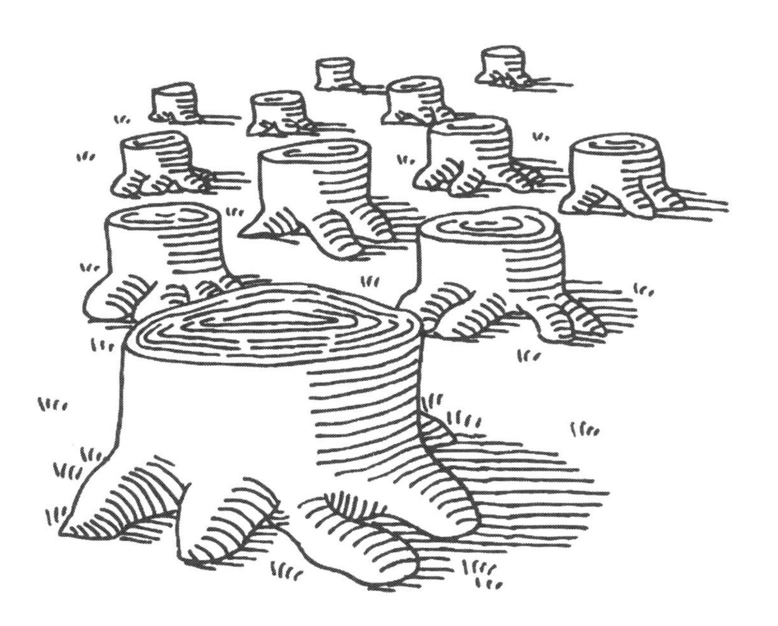

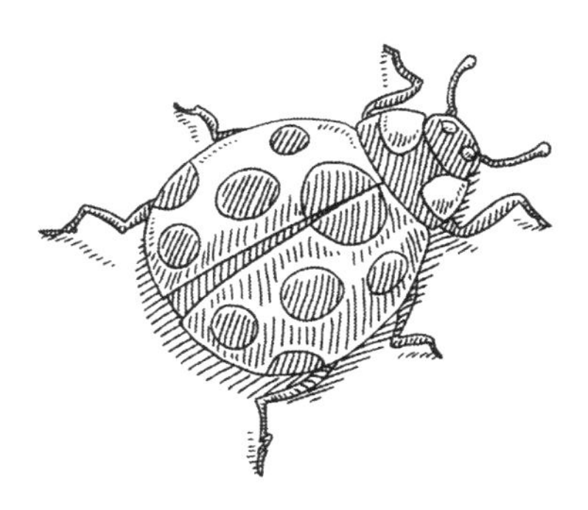

CITIZEN SCIENTIST
by **Shirley Smith Duke**

Scoop up your sweep net.
Come and join me.
We'll search for ladybugs,
rare ones that hide.

Look for the patterns
you'll see on their backs.
Examine them carefully—
nine spots or two?

We'll pull out the camera,
record what we see,
upload our images
to send right off—

The Lost Ladybug Project
needs citizen scientists
and our neighborhood data
in field notes, too,

that we'll write down
for our every find.
Add where and how,
the date, time, and weather.

We're citizen scientists!
We're working together
with other scientists
all over the world!

CLASSROOM IN THE MEADOW

by **Jeannine Atkins**

Children crawl through thickets
to help their father, Professor Fabre.
Their quick hands catch crickets,
carefully tuck bugs in old snail shells.

The children crouch over mules' footprints, small
rain-filled pools, and watch mayflies swarm.
"Observe," Papa says. "Patience shows all!"
They learn that butterflies taste leaves with their feet.

They measure the leaping of fleas
and the vaulting of grasshoppers.
Listen! Crickets hear with their knees
and chirp with their wings.

The children run through fields with cheers for creatures
who are hunters, builders, weavers, miners, architects,
and engineers. Insects are also teachers,
giving lessons in palms, puddles, and mid-air.

Note: Jean-Henri Fabre (1823-1915) is considered a founder of modern entomology, the study of insects. When his seven children grew up, one son continued to work with him, while those who left for other villages or cities in France sent home insect specimens.

SCIENTIFIC INQUIRY

by **Susan Blackaby**

Scientists are like explorers,
using what they know and see
to blaze a trail that, step by step,
will lead to new discoveries:

Formulate, distill and focus,
narrow down, define the gist,
determine scope and pinpoint locus—
this is your **hypothesis**.

Gather all the stuff you need,
to put in play the machinations.
Document the happenings—
these comprise your **observations**.

What things change and what things stick?
Record the outcomes and effects.
Don't presume and don't predict—
collect the **data**: just the facts.

Combine the concrete things you see
with what you know and trials you test.
Interpretation is the key—
results are where you end the quest.

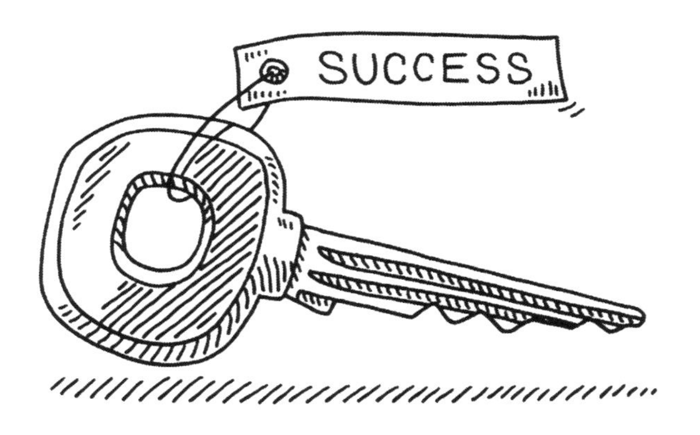

LORENZO LISZT, NON-SCIENTIST

by **Kenn Nesbitt**

Lorenzo Liszt, non-scientist,
researches things that don't exist.
He looks for fur from fish and frogs
and scales that came from cats and dogs.

He hunts for things like hamster wings
and walruses with wedding rings.
He analyzes famous flies
and speculates on oysters' eyes.

He follows leads on pleasing weeds,
investigates banana seeds,
inspects the clothes of ocean crows,
and looks at legless lizards' toes.

He contemplates the common traits
of rattlesnakes on roller skates,
and then explores for dinosaurs
who shop in corner grocery stores.

He thinks about the desert trout.
He studies underwater drought.
He ponders how the purple cow
remained unnoticed up till now.

He scans the skies for flying pies
and tests for turtles wearing ties
and bears who buzz and beep because . . .
well, this is what Lorenzo does.

Although we feel that he should deal
with something that's a bit more real,
Lorenzo Liszt just can't resist
researching things that don't exist.

DiNOS iN THE LABORATORY

by **Kristy Dempsey**

I present to you, my esteemed guest,
a theory putting all else to rest,
a scientific extinction story:
Dinos in the Laboratory.
As archeologists dig down down down,
some curious items have NEVER been found,
perhaps providing us a clue
why dinos paid their mortal due:

√ No safety glasses on their eyes
√ No gloves or aprons, dino-size
√ No rules displayed on any wall
√ No regulations to follow at all
√ No fire extinguisher on hand
√ No first aid kit for quick demand

It's plain to see in science class,
these dinos surely DID NOT PASS!
Did science make them go extinct?
I might be wrong. What do you think?
If you have your own suspicions,
don't recreate these dino conditions.
Take my advice. Expect the worst.
Always remember, safety first!

THINGS TO DO IN SCIENCE CLASS

by **Laura Purdie Salas**

Look at labels.
Ask advice.
Be sure to check directions twice! Wear

Solid shoes to shield your feet,
And keep your table clean and neat.
Follow rules that you are given.
Explore
The startling world
You live in.

SUPERHERO SCIENTIST

by **Joan Bransfield Graham**

As a Superhero Scientist,
I need some special gear.

I wash my hands, put on gloves—
safety goggles always near,

right next to my microscope,
on a handy shelf.

If I want to save the world,
I have to save myself.

LAB TIME

by **Renée M. LaTulippe**

Goggles? On.
Gloves? Got 'em.

Workspace spotless
top to bottom!

WELCOME TO THE SCIENCE LAB

by **Heidi Bee Roemer**

Please quietly enter
the science lab center.
Remember to follow the rules.

Lab work is awesome,
but always use caution
with chemicals, burners, and tools.

You'll be given a pair
of goggles to wear,
plus gloves, a lab coat or apron.

As you measure and pour
always be sure
to wipe away spills at your station.

You may whip up a brew
of some gloppity goo;
Never act on the foolish notion

of tasting the stuff—
you may go belly up!
It might be a poisonous potion.

If you make a mistake
and a beaker should break,
bring it straight to your teacher's attention.

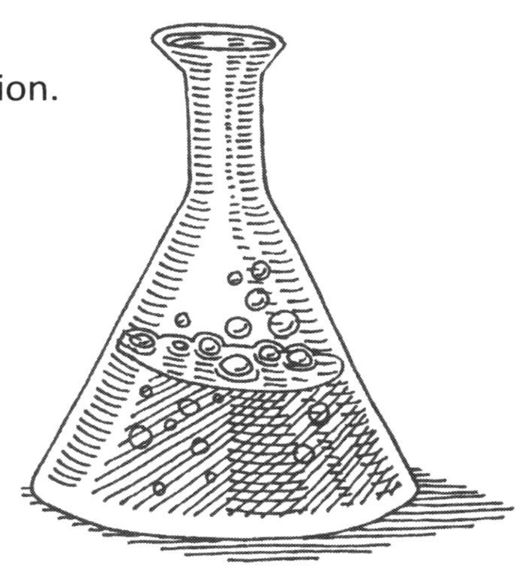

But don't mess around
or you may be found
on the receiving end of a detention.

What a success!
You've proven your test.
Time to put your equipment away.

Wash your hands. Now you're through.
You've learned something new—
and for lab safety you've earned an A!

I HAVE A QUESTION

by **Anastasia Suen**

I have a question,
a question,
a question.
I have a question.
I need to know why.

I look for the answer,
the answer,
the answer.
I look for the answer.
I give it a try.

QUESTIONS, QUESTIONS

by **Ann Whitford Paul**

How does this phone,
smaller than my hand,
thin as a candy bar,
carry conversations
through space
back and forth
and back again?
Why, with miles between us
and no tube,
no wire, connecting us
we can still hear
loud and clear?
How is it possible
we can talk
for hours . . . hours
without other conversations
jumbling into ours?
Questions, questions!
Where are the answers?
I need to find them all.

LATE NIGHT SCIENCE QUESTIONS

by **Greg Pincus**

Do sneakers make me fast?
How long does winter last?
Is goop the same as goo?
What can't a robot do?
What makes a motor go?
Can we drink H_30?
Why is the ocean deep?
Why do I have to sleep?

BACKWARDS

by **Janet Wong**

Everyone is asking WHY.
Why is the sun hot?

I know I should wonder why but
I'm thinking:
Well, why NOT?

INQUIRY

by **Cynthia Cotten**

There are so many questions
I could ask.
Which one
is the right one?

Some questions are simple—
Who?
What?
Where?
When?

But simple questions usually have
simple answers.
They take me only so far.

There are others, though—
Why . . . ?
What if . . . ?
Could I try . . . ?
Can I test?
Can I measure?
Can I describe?

These questions
lead to observations—
to answers that go deeper,
further,
that might even lead to
more questions.

I've heard that no question
is wrong
if you don't know the answer.
I guess the right one
depends on what you want
to know.

WIKI ALERT

by **Debbie Levy**

Oh, those tricky wikis!
Rounding up wisdom from hither and yon,
Hither says *pro*,
Yon argues *con*,
A tower of facts—
An info-nomenon.

But what is real?
What is true?
Is the truth up to you?
'Midst the cons and the pros
Is there someone who knows?
Could be. I suppose.

To tease out the truth in a wiki
You've got to be wary and picky.

TELL IT TO THE COURT

by **Janet Wong**

I didn't mean to copy.
I would never plagiarize.
The words
somehow just . . .
went
straight from the screen
into my eyes
and down
into my fingers
and into my report.

Please-oh-please
don't tell me
to "tell it to the court"!

SINK OR FLOAT
by **Janet Wong**

Popsicle stick.
Crayon. Straw.
Brand new shiny penny.

I'm guessing which
will float like a boat—
I wonder which will. Any?

Take a guess:
what do you think?
Try it—check if it will sink!

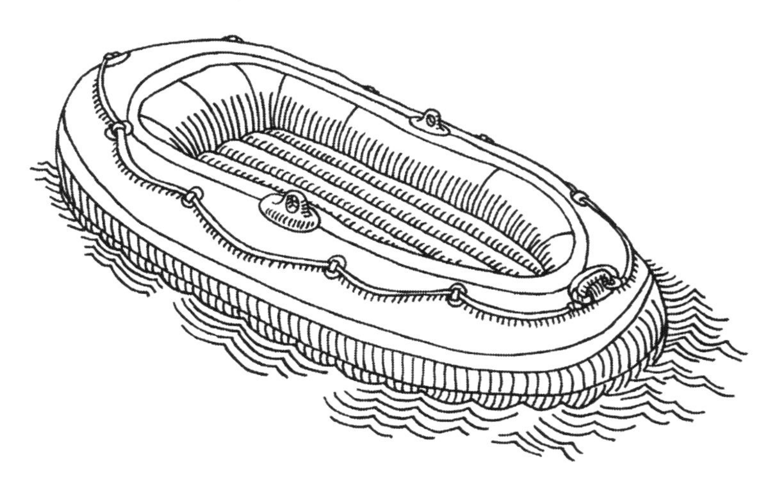

WHICH ONES WILL FLOAT?
by **Eric Ode**

Which ones will float?
Which ones will sink?
Which will grow larger?
Which ones will shrink?
We'll test and investigate,
watch and compare.
But will we agree
with the answers we share?
Maybe we won't,
and then, if we don't,
we'll try, try again
and decide what to think.
Which ones will float?
Which ones will sink?

Which will turn purple?
Which will turn pink?
Which will smell pleasant?
Which ones will stink?
We'll study the data
and enter our claim.
Maybe our answers
will look just the same.
But maybe they won't,
and then, if they don't,
we'll try, try again
and decide what to think.
Which will turn purple?
Which will turn pink?

TESTING MY HYPOTHESIS

by **Leslie Bulion**

I'm testing my hypothesis:
Cats love the color red.
I think this is the reason that
My cat sleeps on my bed.

I set up an experiment to test it:
False or true?
I swap my clean red blanket for my brother's—
night-sky blue.

I feed my cat and scratch her ears.
My tooth-brushing is done.
I'm undercover, snug in blue.
Experiment's begun!

When morning comes,
My furball's at the bottom of my bed.
"No cat hair here," my brother calls,
"She didn't choose the red."

My cat bats at my wiggling toes.
She sniffs them through the sheet.
The next hypothesis I'll test?
My cat loves smelly feet!

A DOG'S HYPOTHESIS: ZOEY'S GUIDE TO GETTING MORE GOODIES

by **Susan Taylor Brown**

If
when my human comes home,
I race to the door and bark and jump
to let her know how happy I am
that she came home to me—
then
I will get a goodie.

Result?
A pat on the head
and not one treat to eat.

If
I stand next to her,
staring at the cookie jar with sad eyes,
letting her know I am starving, wasting away,
letting her know I will soon be an invisible dog—
then
I will get a goodie.

Result?
One tiny treat,
barely enough to taste.

If
I whine, just a little,
then fetch a toy when she asks—
if I sit and lie down,
if I even play dead
(I am good at pretending)—
then—

Result?
Jackpot!
Little liver cookies
fall on the floor
all around me.

TESTING MY MAGNET

by **Julie Larios**

Flowers? No. Dirt? No.
Socks? No. Shirt? No.
Hamster? No. Snake? No.
Plastic scoop and rake? No.
Glue? Paint? Paper? Clay?
Sneakers that I wore today?
No, no, no, no . . .

Pile of metal paper clips—
Yes! Hooray for paper clips!
Shiny whistle? Metal fan?
Dented metal garbage can?
Hammer head, bag of nails?
Ring of keys? Rusty pails?
Yes, yes, yes, and yes!

Results of my experiment?
Magnets are mag-nificent!

DESIGNING AN EXPERIMENT: WILL A CAR ROLL FASTER DOWN A STEEPER SLANT?

by **Avis Harley**

Changing one letter at a time, you can turn *"slant"* into *"speed"*—

SLANT
S**C**ANT
SC**E**NT
S**P**ENT
SPEN**D**
SPE**E**D

Changing one variable at a time, you can see *"slant"* turn into *"speed"*—

SCIENTIST AT WORK

To test car speed on the slant of a slope,
too many variables leaves you scant hope!
But you're on the right scent if you change one by one:
time well spent in experiment fun.
Will the car roll faster? Spend a while on this quest.
Each variable for speed needs its own test.

Note: This poem is a Doublet, a form based on a word game often appearing in newspapers and magazines for children: turning one word into another, one letter at a time. Avis Harley was probably the first to turn this game into a published poem (in her book *Fly With Poetry: An ABC of Poetry*, where she changed "sleep" into "dream").

MY EXPERIMENT

by **Julie Larios**

I tried each possibility,
I tried it all, I tried my best,
I tried to think, I tried to see,
I tried things out, I didn't rest,
I thought I had it, I thought I knew,
I thought what I had was good and true,
but the bottom caved in, the top spilled out,
I couldn't figure the darn thing out,
it all collapsed, it all fell down,
the smile I smiled became a frown.
I didn't succeed, so tomorrow is when
I have to try and try again.
That's good advice, that's right, I guess—
but meanwhile (*sigh) what an awful mess.

PAPER AIRPLANES

by **Janet Wong**

Each team in our class has twenty minutes
to make a paper plane that can fly the farthest.

One sheet of paper per plane.
No other stuff.
Five pieces of paper per team for models.
Each team works in a separate area. No spies.

We brainstorm.
Short and wide or long and thin?
Wing tips up or down or flat?
Pointed nose or squared off?

We make five models and test them all.
With one minute to choose our favorite,
our best plane flies straight into a wall
on its third test flight at the very same time
that our principal walks through the hallway
and steps on it. *Crunch!*
It is broken beyond repair.
Glenn crumples it into a ball and throws it.
It goes farther than anything else we made.
We have ten seconds left when—*ding dong!*—
a question pops into our minds.
A stupid question?
Maybe. But we run to ask our teacher anyway:
Does it have to look like a regular plane?

Kids laugh when they see our ball-plane.
But no one laughs when we jump and shout:
We won! We won!! We won!!!

HAWKING TIME

by **James Carter**

Stephen Hawking
hear my rhyme.
Tell me what
you think of time.

Is it straight?
Does it bend?
When it's over—
well, what then?

Was there time
when time was not?
Who winds up
the cosmic clock?

Mr. Hawking,
if you please,
do solve all
these mysteries.

I know what you
might say to me—
"Go down to
the library!"

DID YOU KNOW?

by **Julie Larios**

. . . it snows metal on Venus?
That's what some scientists say.
One thing I wouldn't want
is to be there on a snowy day.

But one thing I *do* want—
one thing I've wished—
is to discover things like that
when I'm a scientist!

CAN OUR EYES FOOL OUR TASTE BUDS?

by **April Halprin Wayland**

He loves the green drink
She prefers red.

Guess what?

They're both the same!
The taste is in their heads!

Note: This poem was inspired by a psychology experiment that uses apple juice and red and green food coloring to examine whether people's perceptions of taste are influenced by their sight. Check out: Education.com/science-fair/article/can-eyes-fool-taste-buds/

CAPILLARY ACTION

by **Joy Acey**

I put my stick of celery
in my cherry drink.
Three days later
the leaves turned pink!

Tell me your reaction.
Tell me what you think.
Could capillary action
happen when I drink?

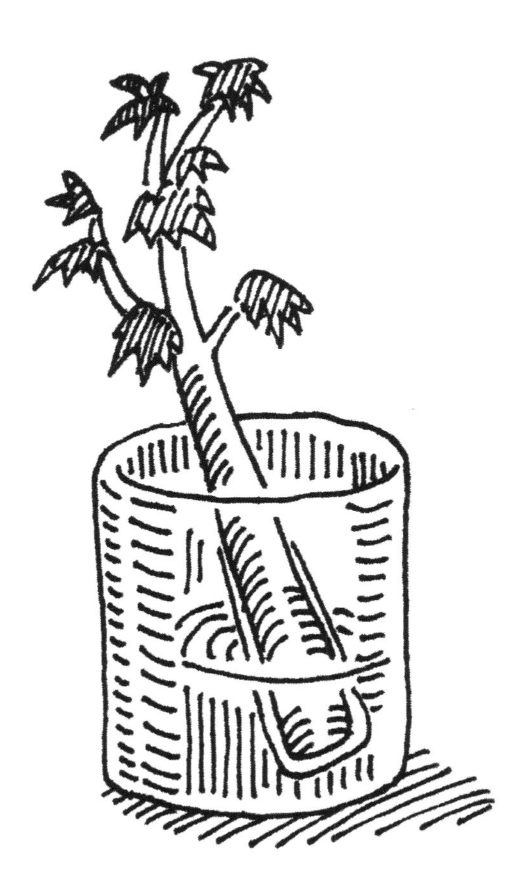

MY PHOTO EXPERIMENT

by **Janet Wong**

For my science fair project
I took three favorite photos.
I copied each file ten times.
Then I played with the settings,
adjusting saturation,
then scribbling down
notes of what I'd done.
I moved onto contrast,
exposure, and tint.
I got so excited
I just had to print.
I stopped taking notes
and I worked the slider bars
and I printed and I printed.
I got pretty far on my project—
I almost got to the end—
when I realized . . .
without notes,
I'm doing it all
AGAIN.

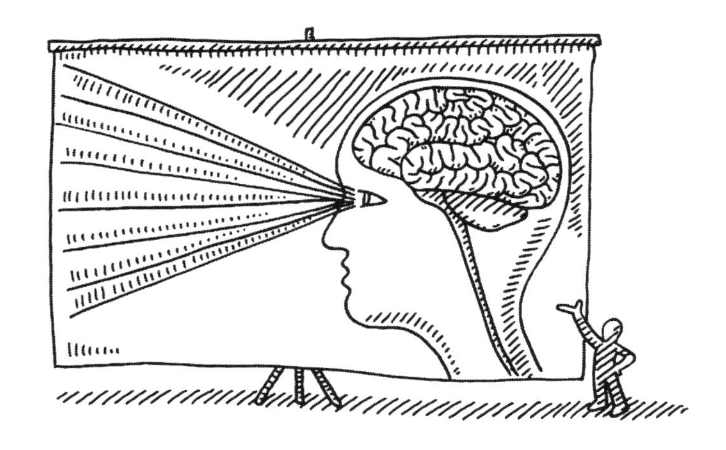

(SUPER)POWER: (TO THE)POINT

by **Kristy Dempsey**

Flash! Shazam! I slide onscreen,
designed to grab attention.
Nothing more and nothing less
than what deserves a mention.
Clear and focused, on the scene,
I'm breaking down this joint,
brandishing my superPower:
Point by Point by Point.

SCIENCE FAIR DAY
by **Eric Ode**

Look at that!
Whoosh!
Splat!
Gurgle.
Pop.
Hooray!
It's Science Fair Day!

SCIENCE PROJECT
by **Lee Wardlaw**

O, Dormant Cone, fear
not! I am Pélé, Goddess
of Fire, who rules your

throat and ire. Awake!
Swallow my potion! Release
what smolders, sickens,

below. You choke. Cough—
burble scarlet froth! My class
erupts in applause.

SCIENCE FAIR

by **Irene Latham**

The graphics
I created and pinned
to the felt board

explain why my eyes
could never be brown,
my hair only blond.

I wonder if Mendel's
theory of genetics
also applies to why

I'm shy
and can speak
to the judges

only in a quavery voice
that betrays my shaky
hands and knees.

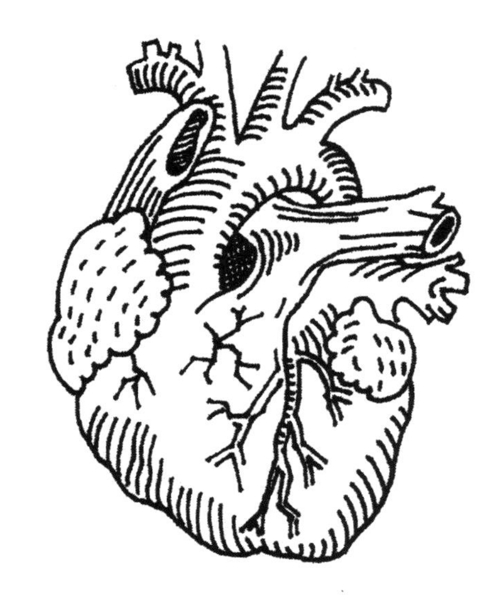

FOR THE SCIENCE FAIR . . .

by **Ann Whitford Paul**

Mom got me a cow heart
from the butcher.
These are my observations.

Color: dark red with
pale yellow fatty areas
Smell: like fresh hamburger meat
Weight: two pounds, one ounce
Texture: smooth
Shape: a blob, that had
to be manipulated into
a valentine heart
Two main arteries: the aorta and the pulmonary
Inside: four chambers, four valves.

These are my observations.
Now I know everything
about the heart,
everything except . . .
the wonder of its beat.

SCIENCE FAIR PROJECT

by **Eric Ode**

I thought I'd win a ribbon
and my work would be rewarded.
My research, clearly catalogued,
my variables, recorded;
I proudly set my project
with the other kids' displays—
the vinegar volcanoes
and the cardboard hamster maze.
I waited for my trophy,
feeling confident and grand.
And that's about the time
when things got slightly out of hand.
Now my teacher's looking troubled,
and I bet she's holding grudges.
My project ate my tri-fold
and then seven of the judges.

MY PROJECT FOR THE SCIENCE FAIR

by **Kenn Nesbitt**

My project for the Science Fair
was positively cool.
I built myself a time machine
and showed it off at school.

Inventing it was not too hard;
I had a little aid.
My future self came back in time
and showed me how they're made.

FROGGY

by **Charles Waters**

Sleeves rolled up, utensils at the ready,
I look down at who we're about to dissect
for science class.
I take a deep breath as
Froggy is sliced open
like a hunk of cheese.
Some kids walk away from this spectacle.
I'm fascinated:
kidneys, lungs, intestines,
gall bladder, pancreas, heart and *liver*
are all connected like a symphony.
"You have these same parts within you."
Mrs. Lance says.
"When it comes to our animal friends,
we're more alike than we are different."
"Wait a minute," I say,
looking into my magnifying glass,
"There's one thing
we don't have in common with Froggy."
"What's that?" Mrs. Lance asks.
"Our last meal before we die won't be
beetles, ladybugs, earwigs and slugs."

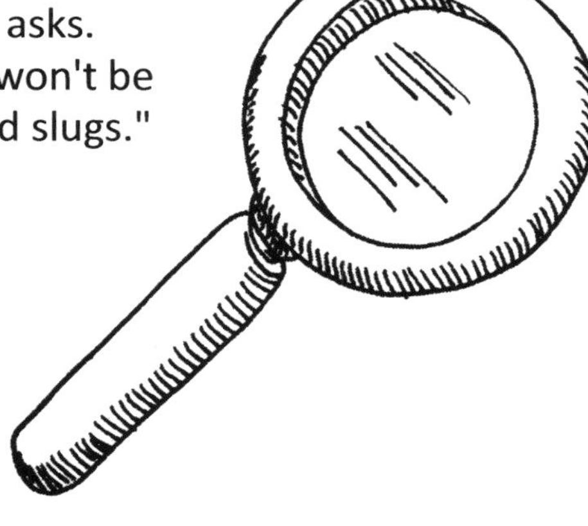

MRS. SEPUKA'S CLASSROOM PET

by **Ken Slesarik**

Who will care for our classroom pet?
I haven't any takers yet.

Take him home for winter break.
It's a hermit crab for goodness sake.

How 'bout you? Would you? Would you?
There's just some things you'll need to do.

He comes complete with "crabitat."
You'll want to put some moss in that.

Needs water too, both fresh and salt.
He likes to climb and move about.

For food he does have certain cravings
like fish heads and coconut shavings.

Lastly please do not forget,
call him Harry not Harriet.

THE CLASS PLANT
by **Janet Wong**

The plant on our bookshelf
is turning yellow, drooping
and dropping leaves.

We talk about what it might need.
I feel like a doctor in a hospital
talking about a sick patient.

Jack wants to give it more water.
Kayla says the soil smells moldy
and feels soggy—too much water.

Chris asks: How about more light?
We move the plant to a bigger pot.
I place it carefully near the window.

I have a prediction: "Tomorrow
our plant will be dancing!"
They laugh but they will see.

The next day the weather is very hot.
The air conditioning is blowing hard
right down on the leaves of our plant.

Hot weather
+ air conditioning
Dancing Plant!

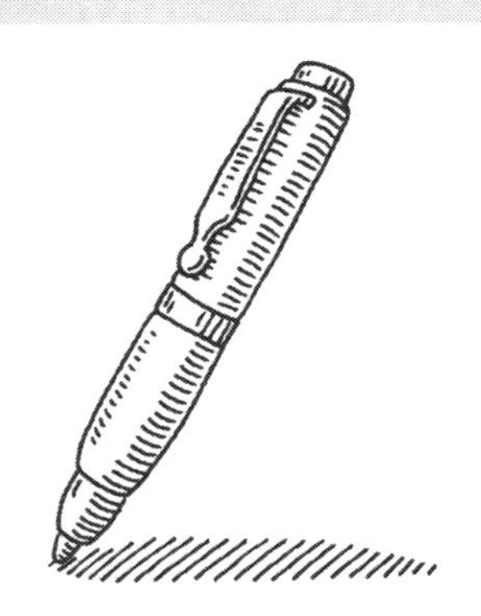

TEACHER'S LOOK
by **Shirley Smith Duke**

"Don't tap on the tables,"
Our teacher said once.
But pens click so nicely
My hand failed to note.

Taps send compressed waves
Right through the air.
The vibrations pass through,
Inside both inner ears.

She stops—Oh, the look!
Now what shall I do?
Transport to a vacuum
Where sound can't move.

She points to her notebook,
And writes down my name.
Then I vow to stop tapping
And fold my hands when . . .

The whole class starts tapping—
A symphony of sound
Fills the room with vibrations
Of more taps and noise.

The synchronized beats
Make her give up her look.
My teacher starts laughing
And closes her book.

MEET MR. WIZARD

by **George Ella Lyon**

No science lab in my
school, no library
even. But Mr. Swift
does experiments
anyway.

My favorite is the egg
and the bottle.
The egg has to be boiled
and peeled, the bottle
empty.

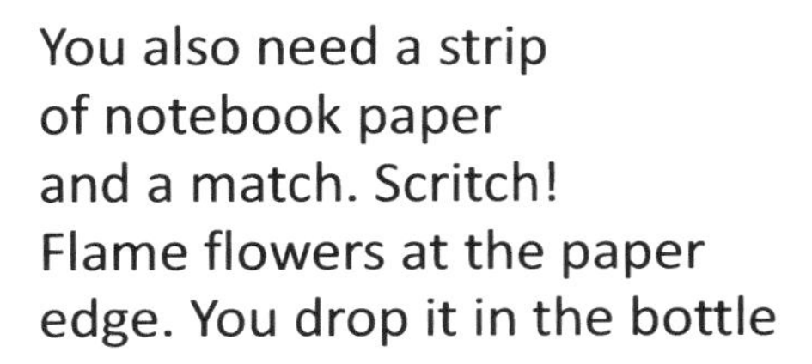

You also need a strip
of notebook paper
and a match. Scritch!
Flame flowers at the paper
edge. You drop it in the bottle

and place the egg on
the bottle's lip, blocking
the air so the flame goes
lower
lower
out.

There's a pause
and THWUP!
The egg
slooks through the neck.
This demonstrates

the vacuum
and proves
that science
can make
you laugh!

PUMPKIN EXPERIMENT

by **Mary Lee Hahn**

We put one pumpkin
in the land lab—
left it there
all fall long.

It shrank and shriveled
in the land lab—
after winter,
it was gone.

Where we left it
in the land lab—
pumpkin plants
are growing green.

Five fat pumpkins
in the land lab—
four to carve
and one for seeds.

Note: Our land lab is an outdoor
laboratory at our school.

FIRST SCIENCE PROJECT

by **Lesléa Newman**

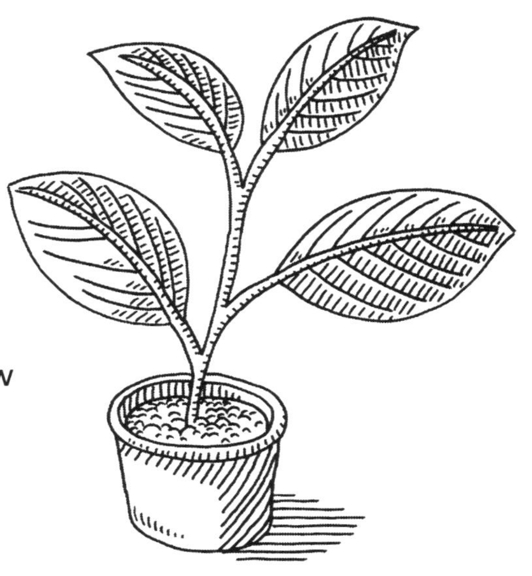

I ate the avocado
That we bought at the store,
And it was so delicious
I wished that I had more.

My mother cleaned the pit up,
And handed it to me.
"Get some toothpicks and we'll grow
An avocado tree."

We pierced the pit with toothpicks
And perched it on a glass,
Then filled the glass with water.
"Now watch what comes to pass."

Soon some roots were dangling down,
Straggly, thin, and white.
Soon we saw a bright green stem
That reached up toward the light.

Soon the plant was growing leaves,
We welcomed every one.
Each leaf so green and shiny,
Unfurling towards the sun.

And when the plant grew bigger,
We planted it in dirt.
The next day it was taller,
An overnight growth spurt!

The plant grew even bigger,
Three feet from stem to root.
And though we watched and waited,
It never did bear fruit.

"It doesn't really matter,"
My mother said to me.
"We still have something lovely:
An avocado tree!"

BREAKFAST ALCHEMY
by **Mary Quattlebaum**

Science in the kitchen!
My little brother claps.
I let him measure flour
(one cup, to be exact).
Next, a spoon of sugar,
one cup of milk, then—crack!—
a single egg, a dash
of oil, and . . .

Flat! Our flapjacks
didn't fatten. Why?
No baking powder! Oops.
A small mistake that we
can mend; we measure, add,
and try again.

My brother stirs and stirs.
Our oopy-goopy glop
grows smooth—and now
the griddle's really hot.
We pour another batch
and soon our liquid, white-ish
dots begin to warm
and thicken, puff and rise . . .
Look now! Oh, wow!

Solid gold upon our plates:
Super-Yummy Science Cakes!

MICROWAVE OVEN
by **Janet Wong**

It takes twenty minutes
to make a batch
of my favorite
blueberry pancakes
from scratch.

On weekdays we need
every minute of sleep
so we microwave
pancakes:
one minute—beep!

But watch out:
the pancakes get HOT
in the middle—
much hotter than when
they come off the griddle.

Mom thinks
microwaves cook
from the inside-out
but that's not true:
I've learned about

how microwaves
get into the food
just a bit and activate
water molecules.
Once hit,

they vibrate
and bounce
to make lots of heat.
Electromagnetic radiation—sweet!

FIVE O'CLOCK RUSH

by **F. Isabel Campoy**

Game's in thirty minutes
and father is late.
How about dinner?
he asks in dismay.

I open the fridge
and take what I saved
two pieces of pizza
to heat
in
the microwave.

Only one minute.
Dinner is served!

PRISAS A LAS CINCO

por **F. Isabel Campoy**

En treinta minutos
empieza el partido.
Papá llegó tarde
—¿Y de cena? —
pregunta afligido.

Abro la nevera,
saco mi sorpresa:
¡dos trozos de pizza!
Rápido las meto
en el microhondas.

Y casi enseguida,
en solo un minuto
¡La cena está servida!

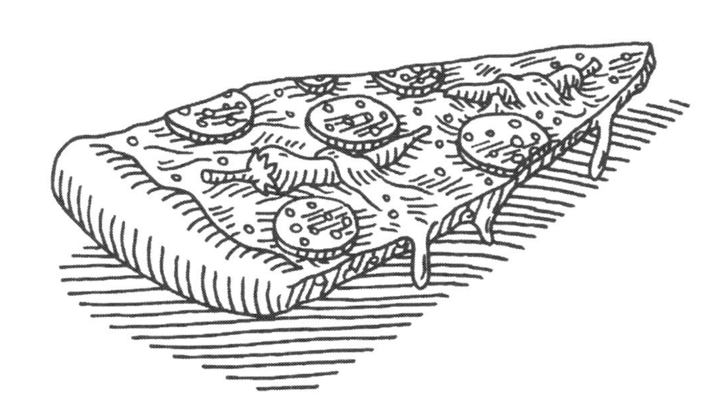

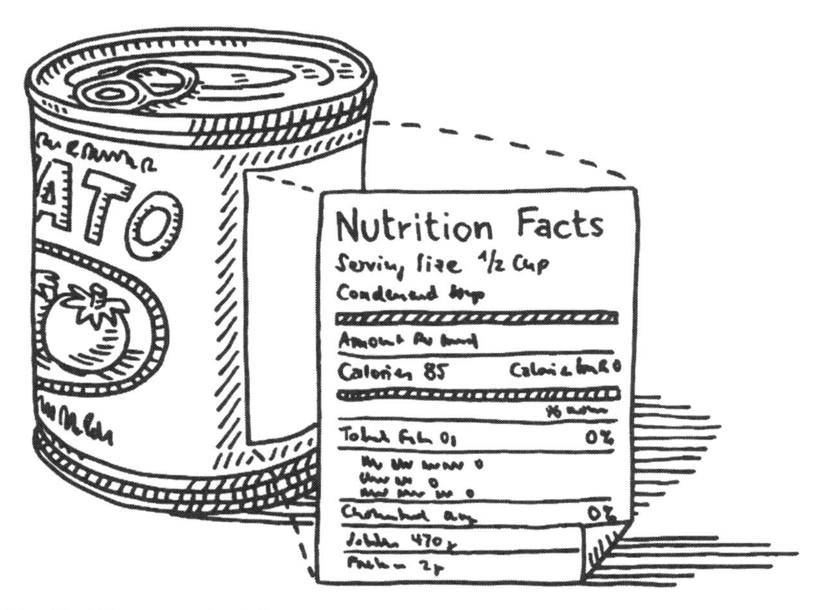

FOOD FOR THOUGHT

by **Robyn Hood Black**

You won't find a character, setting or plot
on the side of the cereal box Dad bought.

But wait! There's still something tasty to read.
The **food label** has information you need.

Ingredients tell you what is inside.
(See sugar and salt? They were trying to hide.)

Your body needs **protein, carbohydrates,** and **fat**.
A good bit of this, just a little of that.

Vitamins help keep you active and strong—
minerals, too, when they tag along.

Check out the **calories** per **serving size**.
Then make a choice that is healthy and wise!

POWER BLENDER AND AHA!

by **F. Isabel Campoy**

Seedless smiles,
juicy hugs,
fresh banana kisses
tickling kiwi laughs.

Very berry sense of humor
mango-tango to sing and dance,
all packed tightly into a blender
with a power button, and aha!

The best smoothie served
to friends and fans!

ENCHUFA LA BATIDORA Y ¡YA ESTÁ!

por **F. Isabel Campoy**

Sonrisas con uvas sin semillas
abrazos muy jugosos,
besitos de plátano fresco
risas de kiwi con cosquillas.

Chistes de fresas y frutillas
mango-tango para cantar y bailar,
todo metido en la batidora
y listo para licuar.

Para mis amigos,
¡el mejor batido está servido!

MOLD

by **Charles Waters**

When food becomes too old
It's visited by mold,
Fuzzy green, blue, black gunk.
Please, throw it out; it's junk.
Although some mold is helpful, yes—
It's hard for you or me to guess
So, please be safe: toss it! Why?
You could get sick from fungi.

WATER + DIRT =

by **Rebecca Kai Dotlich**

Mud.
I will make some for us.
Easy! No fuss.
A handful of dirt.
Some water—squirt!
Need one small stick.
Stir until thick.
What does mud make?
A puddle.
A cake.
And goodness knows,
it feels good between toes.

OUR TRUCK

by **Janet Wong**

Last year
our truck
got busted up—
and we couldn't fix it.
Now it's covered in rust.
Dad says it's just
water and air
mixing with
the iron in the steel.
And the paint
that used to be so smooth
is now as bumpy
as an orange peel!

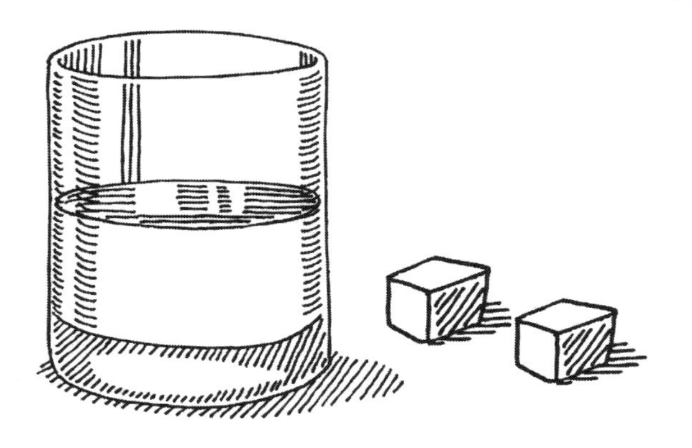

TAKE BACKS

by **Janet Wong**

Some things you can take back
without any trouble—
like sand mixed with rocks
or peas mixed with rice.

Some things you can't take back—
like salt in cake batter.
Grandpa thought
he grabbed the sugar but
he put a cup of salt
in by accident—

no taking that back, for sure—
especially after the batter's been baked!

Want a slice of Salty Chocolate Cake?

SUGAR WATER

by **Janet Wong**

We dump some lumps of sugar
in cold water. Stir, stir, stir.

There's sugar at the bottom
though we stir, stir, stir, stir, stir.

What happens with hot water?
The sugar disappears!

Excuse me, Sugar:
are you still here?

CHANGES

by **Janet Wong**

Physical change:
your popsicle starts to melt
in the sun.
Pop the popsicle
back in the freezer.
The change can be undone.

Chemical change:
frying a burger
you burn the bottom black.
With a chemical change
unfortunately
there's no going back.

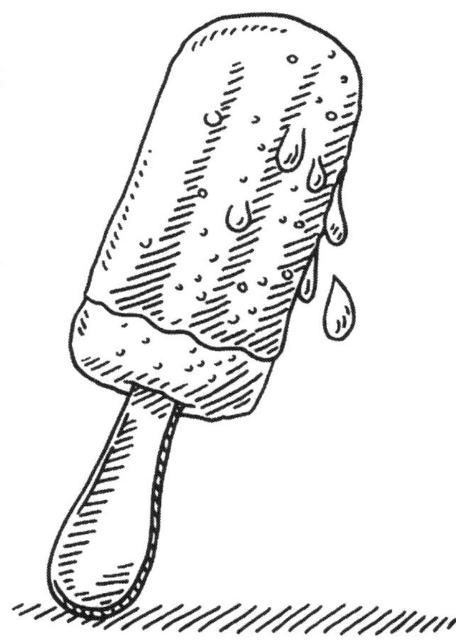

THE BRINK
by **Janet Wong**

I fill a cup to the top
with crushed ice,
pour juice to the brim,
neat and nice.
Mom thinks
it's on the brink of disaster.
When I take just a sip,
she shouts: "Drink faster!"
When the ice melts,
will my drink spill out?
I think there's nothing
to worry about
but I wait and I watch.
The ice seems to shrink.
PHEW! Okay—
time to drink!

LIQUIDS CAN'T CONTAIN THEMSELVES
by **Heidi Bee Roemer**

Sticky honey leaks from a jar.
Oozy ketchup squirts too far.
Hot soup overfills its bowl;
Liquids dribble and ripple and roll!

A bucket, a jug, a jar, or a vase
keeps each liquid in its place.
But liquids cannot keep a shape;
they're always seeking to escape.

Oops! You spilled your carton of juice.
Aren't liquids messy when they get loose?

QUESTIONS THAT MATTER
by **Heidi Bee Roemer**

What is a solid?

"I am," says the wall.
"My size and shape remain the same;
I don't change at all."

What is a liquid?

"I am," says the milk.
"My carton gives me shape.
I'm a puddle when I'm spilt."

What is a gas?

"I am! Call me Steam-y!
My vapors fill the room,
but you probably can't see me."

I WILL BE A CHEMIST: MARIO JOSÉ MOLINA

by **Alma Flor Ada**

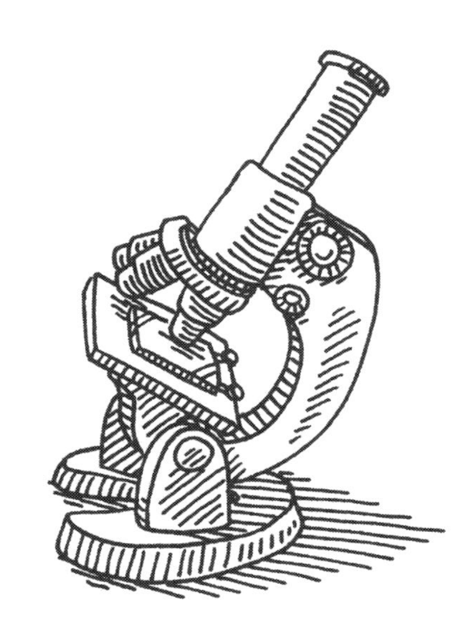

Only a drop of water
but looking under the microscope
I see things that move inside that very drop.
My aunt Esther has given me a chemistry set.
She says that everything—water, air, earth,
the trunks of trees and our own skin—
is made of small particles we cannot see.
She explains that even these molecules
are made of chemical elements;
just around a hundred elements
combine to make all that exists.
I have started today
in my simple lab in the old unused bathroom
to study these elements.
I will know the secrets of the universe.

Note: Mario José Molina, born in Mexico in 1943, and a resident of Mexico and California, won the Nobel Prize in 1995 for his contribution to understanding the ozone layer. He continues his work to solve pollution problems and to study atmospheric particles and their effect on clouds and climate.

VOY A SER QUÍMICO: MARIO JOSÉ MOLINA

por **Alma Flor Ada**

Sólo una gota de agua
pero mirándola con el microscopio
veo cosas moviéndose en ella.
Tía Esther me ha regalado un juego de química.
Dice que todo—el agua, el aire, la tierra,
los troncos de los árboles y hasta nuestra piel—
está hecho de pequeñas partículas que no podemos ver.
Explica que las moléculas
están formadas por elementos químicos.
Apenas unos cien elementos
se combinan para crear todo lo que existe.
He empezado hoy
en mi simple laboratorio en el viejo cuarto de baño
a estudiar estos elementos.
Voy a saber los secretos del universo.

Nota: Mario José Molina, nació en México en 1943 y ha vivido en México y en California. En 1955 recibió el Premio Nobel por ayudar a comprender la capa de ozono. Continúa estudiando cómo resolver problemas de polución y el efecto de las partículas atmosférica en las nubes y el clima.

ELEMENTAL

by **Jeannine Atkins**

Marguerite Perey washed test tubes that once
glowed blue. Could she discover an element?
Madame Curie, who ran the laboratory,
had found two: polonium and radium.

Most elements had already been found.
Calcium is part of bones, pearls, and chalk.
Iron is both in blood and far underground.
Oxygen and hydrogen are almost everywhere.

Would another element glow like gold,
last like lead, stink like sulfur, or glare like neon?
Every element is unique, we're told.
Arsenic poisons. Chlorine cleans.

Marguerite isolated francium, the last element
to be found in nature. Or will there be more?
Seekers and celebrants raise beakers
to all that's beyond and what came before!

Note: Marguerite Perey (1909- 1975) worked in Paris with Marie Curie, who won two Nobel Prizes, and her daughter, Irène Joliot-Curie, who won one Nobel Prize. Both encouraged the work of talented female chemists and physicists, but it wasn't until 1962 that Marguerite Perey became the first woman to be admitted to the French Academy of Sciences.

THINK OF AN ATOM

by **Buffy Silverman**

Think of an atom
so tiny, so small—
a speck of the world
a speck of us all,
a speck of the ocean
a speck of a fly
a speck of a mountain,
a book or the sky.

Imagine that speck
growing wide, growing tall
an atom as large as
your school or the mall.

The atom looks empty—
almost nothing at all,
but there in the center
a tiny tight ball
of neutrons and protons
with mass and with weight.
How many of each?
(for oxygen: eight!)

Its charges are balanced:
a proton adds one,
—(each electron's a minus)
the neutrons add none.

Outside of the nucleus—
that tight little ball—
the electrons are swirling
they're smaller than small
like pieces of dust
whizzing through space
a cloud of electrons
in a zip-zapping race.

An atom is tiny—
astoundingly small—
Trillions lie here
on this dot that I scrawl.

ALBERT EINSTEIN

by **Julie Larios**

Einstein's mind
matched his hair.
Both had something
energetic blow through them
at the speed of light squared.
In fact Einstein's massive frizz—
brain-wise, hair-wise—
was absolutely atomic.

DISCOVERY

by **X. J. Kennedy**

Benjamin Franklin from
Old Philadelphia
Sent up a door key tied fast to a kite,
Guessing that lightning was sheer electricity.
Turned out that wise guy
Was perfectly right.

IMAGINE SMALL

by **Eileen Spinelli**

Imagine something very small:
a rubber duck, a ping-pong ball.

Imagine something smaller yet:
a pebble or a violet.

Go smaller now: a silver bead,
a baby's tooth, a pumpkin seed.

Keep going—
freckle, flea, or gnat,
a speck of dander from the cat.
Imagine that.

And then imagine this—so cool!—
a teeny-tiny molecule.
So teeny-tiny you and I
can't see it with the naked eye.

To think of it gives me a chill.
But there is something smaller still:
the atom!

Billions fit in a fleck of foam
or on the dot at the end of this poem.

Billions.

WHAT CAN YOU MAKE FROM CARBON?

by **Laura Purdie Salas**

Charcoal
Makes drawings and fire

Graphite
Makes words that inspire

Diamond
Makes drill bits and rings

Carbon
Makes all living things

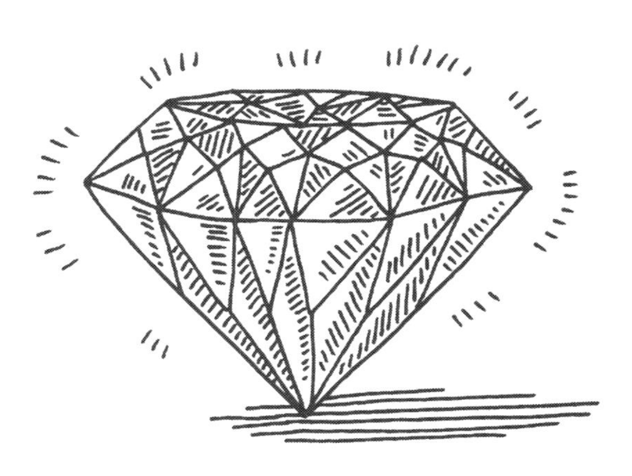

AFTER I MADE A HUGE MESS WITH MY CHEMISTRY SET

by **Mary Lee Hahn**

My dad is mad.
He's really ticked.
My punishment?
To move the bricks

he's saving for paving
a path in the garden.
My only tool?
My little red wagon.

I pile it high.
I push and I pull.
My realization?
My wagon's too full.

I take a break
and think it through.
My decision?
Remove just two.

I guess and test
'till my load's just right.
My success?
Not too heavy, not too light.

My dad is sad.
He heaves a sigh.
My dreadful chore?
More fun with science!

PUSH POWER

by **Janet Wong**

I pull with my hands.
My wagon is stuck.
I push harder with legs.
This time I'm in luck.
My wagon gets out
of the mud
but—
wait!

It zooms
down the hill
straight into the lake!

WINDFALL IN THE ANDREWS FOREST
by **Joseph Bruchac**

The way the giant Douglas fir
leaned after five centuries
showed the way
wind wanted it to go.

Wide roots, spread
into the soil like hands
were not enough to hold.

It crashed down through the canopy,
scattered branches over the stream,
needles and Old Man's Beard lichen
fluttering down like green rain.

The small trees below,
yews, hemlocks, and alders,
were not net enough to slow it.

But the earth and its stones were stronger,
for when the tree struck,
its great trunk broke,
its bark was shed like an overcoat,
and its layers of growth split to splinters.

FRICTION
by **Sara Holbrook**

Speed bumps in the parking lot.
Gravel under my wheel.
Brakes on a subway train
screeching out a squeal.
A zombie dragging a ball and chain.
My eraser tearing at paper.
My father's weekend beard
on Monday pulling at his razor.
A thumb against a finger
when it
makes a snapping sound.
Whatever takes off in a hurry,
friction slows it down.

THANK YOU, ISAAC NEWTON

by **Eileen Spinelli**

My bookshelf falls upon the bed.
Harry Potter bonks my head.
Spaghetti slips—splat!—to the floor.
Clean-up is a messy chore.
Orange juice spills. Socks slide down.
Hail stones ping all over town.
Acorns plunk—ouch!—from a tree.
Oh, the joys of gravity!

GRAVITY

by **Joyce Sidman**

Think of the Earth
as a mama
with a warm, heavy heart.
She's lonely in space.
She reaches out her great arms
and holds us to her:
rocks, trees, elephants,
clouds, kites, air.
We can fly away—of course!
But only so far
before she calls us back.
We can jump
and vault and bounce and twirl;
but always, always,
we return to her.
She worries about growing
older, smaller,
weaker—
like her bleak sister,
the moon.

She holds on tight.
Her hug
encircles the world.

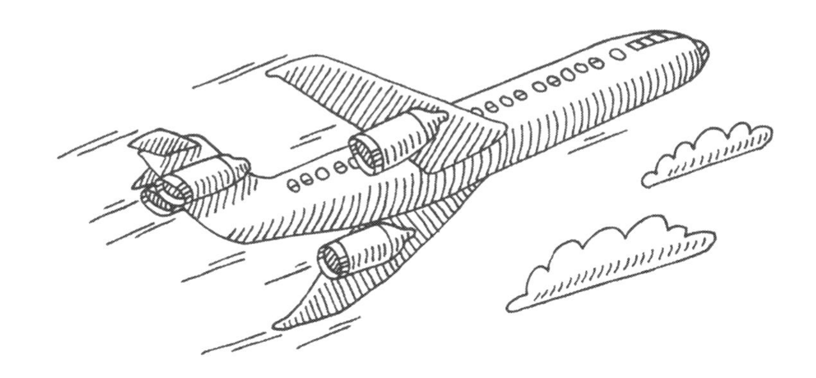

LIFT

by **Marilyn Singer**

When you don't have wings
and feathers, just arms and skin,
gravity will win.

But people, craving
panoramic, learned what's
aerodynamic.

So, with metal, fuel,
invention, victory's
ours: ascension!

ROLLER COASTER RIDE

by **Patricia Hubbell**

We're fastened in and up we go—
To reach the top we're starting slow—

Whoooshhh!

We're diving! Veering!
Climbing! Swooping!
Dropping! Twisting!
Curving! Looping!

Up! Down! Around! Around!
Down! Down! Down!
We're zooming fast! We're racing faster!
Faster! Faster! Faster! Faster!

We're yelling! Screaming!
Shouting! Laughing!
Hooting! Hollering!
Shrieking! Gasping!

What a crazy, great sensation—
All because of . . . ACCELERATION!

GO FLY A KITE

by **Laura Purdie Salas**

Above the kite, the pressure's low.
The air's a streaming, breezy flow.

Below the kite, the pressure's higher.
Up! Up! Up! This one's a fly-er!

Lift versus drag.
Lift wins!
That's why . . .

your kite
breaks
free
and
climbs
the
sky!

FRISBEE

by **Glenn Schroeder**

Why does a Frisbee go far with a fling?
Because it moves air like an airplane wing.

Why does a Frisbee fly flat and not wiggle?
Because it is spinning, which stops the jiggle.

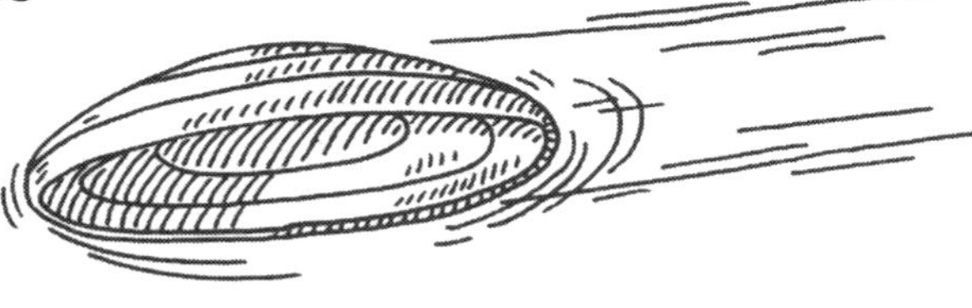

Note: Frisbee is a registered trademark of Wham-O.

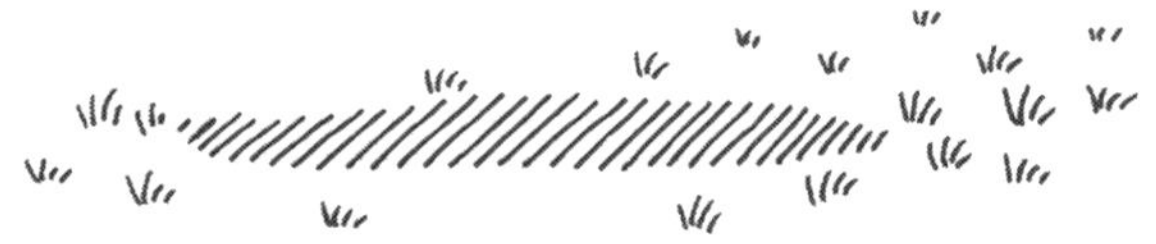

LOVE NOTE TO A MAGNET

by **Patricia Hubbell**

Dear Magnet,
I'm drawn to you.
Irresistible you!
You're so attractive.
You make me feel quite *active*!
And though I know I should shrug,
I can't resist the tug of this feeling of love.
I want to hop, jump, and skip,
Really let rip . . .
So I'm on my big trip—
Past pens, pencils, and papers,
Rulers, tape, and erasers—
Because . . .
I'm drawn to you!
Hugs,
Paper Clip

NO PENGUINS HERE

by **Michael Salinger**

It has been said
that opposites can attract
and while this may not be a fact
between your brother or sister and you
it most certainly will hold true
when magnets are the topic.

Every magnet has two poles—
one north and one south at either end.
Whether two magnets repel or attract
depends on how these poles align.
You'll get the same result every single time—
but go ahead and try it.

South pole to south pole
or north to north
push each other away
but north and south come together—
they stick and they stay.
Because at each pole the magnetic field
is at its max
doing its job, staying on track
ensuring that—at least with magnets
opposites do attract.

PLAYGROUND PHYSICS

by **Jeannine Atkins**

Boys on the seesaw learn about levers.
A girl on a swing pumps her legs, jumps, and lands.
Her sneakers slip and skid. She's grateful for gravity
and friction, which makes her sneakers stop in sand.

Mr. Newton's students play with a bat, a ball, force,
energy, and motion. The pitcher throws balls that spin.
One batter strikes out. The next sees the pitcher twist
his wrist. The batter swings, then sprints and grins.

Quieter scientists watch ants use their jaws
like levers, lifting long blades of grass.
Could ants make a seesaw? There's no time to ask.
Recess is over. The bell rings. Back to class!

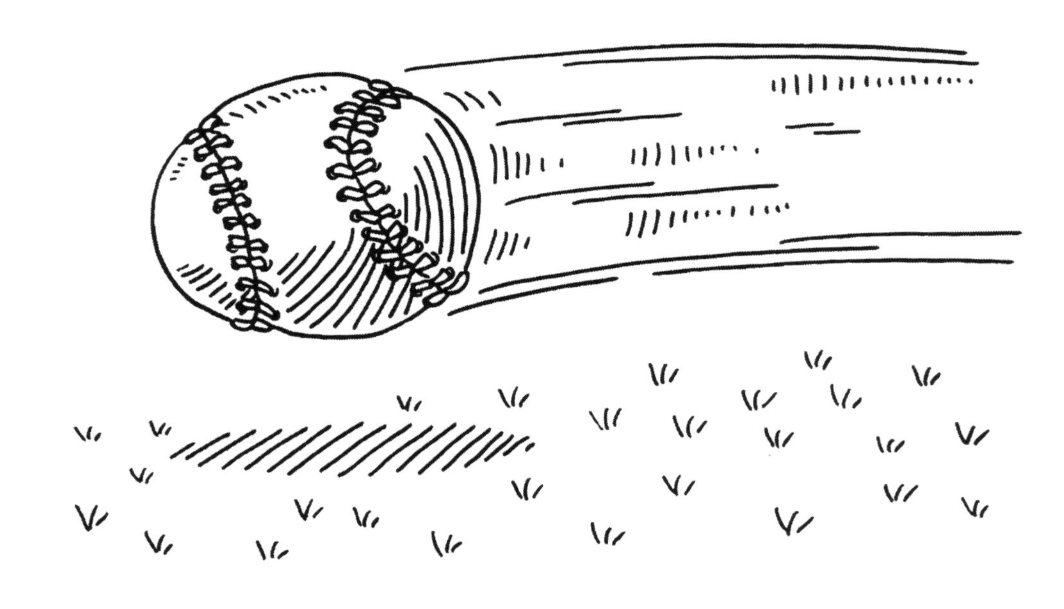

AT THE SPEED OF LIGHT

by **Shirley Smith Duke**

The sun gives off waves of energy,
hurled at the speed of light.
Particles from these waves
pass through the vacuum of space.
Waves are spaced with crests and troughs,
each kind with its own length.
Line up each one to see
waves of the electromagnetic spectrum.

Radios and phones tune in long **radio** waves.
Microwaves heat water in our food.
Infrared warms us in fires and sun's rays.
Visible light waves bring bright colors to our eyes.
Ultraviolet waves burn skin to a sore sunburn.
Cavities and broken bones show in short **X-ray** waves.
Shortest **gamma** waves arrive with lightning flashes.

Long or short, visible or invisible,
moving in different frequencies,
some reaching Earth while others are absorbed
by the protective atmosphere:
in differing wavelengths, varied in size, all
radiant energy.

WHAT AM I?

by **Esther Hershenhorn**

I was born long ago,
in 1956.
Zenith's Dr. Robert Adler gave me my *clicks.*

I fit inside your hand,
I turn ON and OFF and scroll.
I raise and lower volume.
I change channels.
I control.

Wirelessly I work from far away.
I give power to machines.
Did you guess *remote*?
Hurray!

HELLO, HELLO!

by **Janet Wong**

There's a button
on Mom's phone.
She presses it and—
poof!—
we're home,
sitting in the living room
with Grandpa asking,
"Where are you?"
Well, we could be
in Mexico
but you would never
ever know.
It feels like
we're down the hall
when we make
a video call!

ALEXANDER GRAHAM BELL

(a clerihew)
by **Avis Harley**

Alexander Graham Bell,
during a creative spell,
put his mind to the grindstone
and invented the telephone.

SOUND WAVES

by **Amy Ludwig VanDerwater**

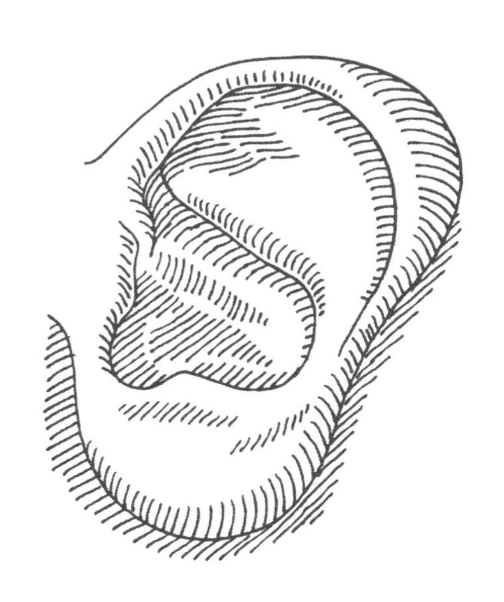

If you have ever seen the ocean
throwing cold waves from her hand
pulling shells from mighty depths
tossing each upon wet sand,
you can understand how sound waves
move like water through dry air.
One-by-one, vibrations follow
pressing sounds from here-to-there.
Sounds can pass through liquids.
Through gases. Solids too.
But sound waves moving through the air
are sound waves meant for you.
Violin or thunderstorm—
each will reach your waiting ear
to play upon a tiny drum.
This is how you hear.

CAN YOU HEAR A CONCH?*

by **Laura Purdie Salas**

A conch looks like a giant ear.
Listen closely—
Can you hear?

A lion's roar?
A gentle snore?
Waves upon the rocky shore?

I had a thought. Can this be true?
Do you think
It hears me, too?

Note: *Conch* is pronounced *conk.*

LiSTEN

by **Amy Ludwig VanDerwater**

I am a sound wave.
I travel through air.
No one can see me
but I'm everywhere.

I vibrate soft whispers
and songs in your ear.
I am a sound wave.
It's me that you hear.

PASS ME THOSE EAR MUFFS

by **Graham Denton**

In our
Science lesson
on sound,
Miss Butters asked us
to name
"one situation
where ear protection
might have to be worn
to save hearing
from being
damaged."

I didn't write
when working
with power tools
or with noisy
machinery
or even
at a
very loud
rock
concert;

I wrote,
"In the classroom,
listening to
Miss Butters shout
when she's really,
really
mad at me
for being so cheeky."

And do you
know what,
after Miss Butters
read that,
I was proved
absolutely
right.

SOUND WAVES

by **Michael Salinger**

Sound travels in waves
like ripples from a penny
dropped in a bucket of water.
Once sound is created
a tick, a tock
a clap
a boom
a whisper
a crackle
a crash, a zoom
its energy rides upon the crest
doing what any wave does best.
Traveling through
air, water, or even a solid
until its energy is finally used up
and the noise fades to quiet
the sound flattened to stillness
not making a peep
until that dog next door barks
setting off another wave
waking you from your sleep.

SOUND WAVES AT BREAKFAST

by **Susan Marie Swanson**

I can hear the garbage truck backing up
beep beep beep,
and then the big roar and ruckus of its work.
Our dog goes crazy *bark bark barking*.
My *clang* baby sister *clang*
bangs her spoon *clang* on her tray.

Shhh.

I'm listening for the vibrations of squirrel claws
scritching up trees.

I like the sound waves from the dog's clinking tags
and his tail thumping on the wooden stoop.

My ears are ready for the pop
when the lid pops off this new jar of jam.

Editors' note: Line 4 in a previous version of this poem used "My neighbor's dog" instead of "Our dog."

TO THE EYE

by **Laura Purdie Salas**

Light is
a jumpy kid
playing hopscotch

bouncing from thing to thing
picking up color like a pebble
to carry in its fingers

lightbulb to French fry to eye

sun to tire swing to eye

campfire to s'mores to eye

always, always to the eye

When light lands, you
don't
hurt
don't
squint
don't
blink
you

see

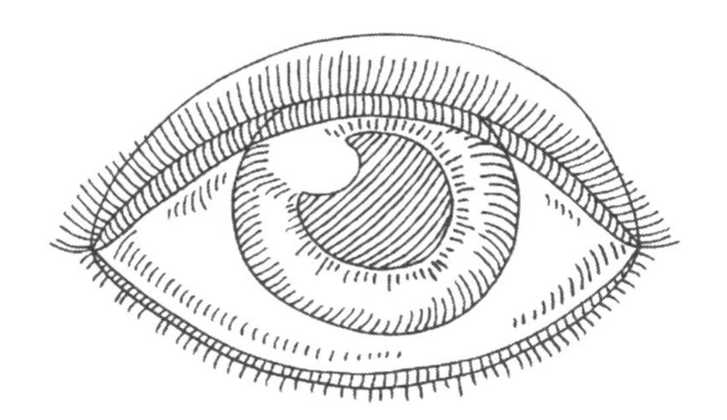

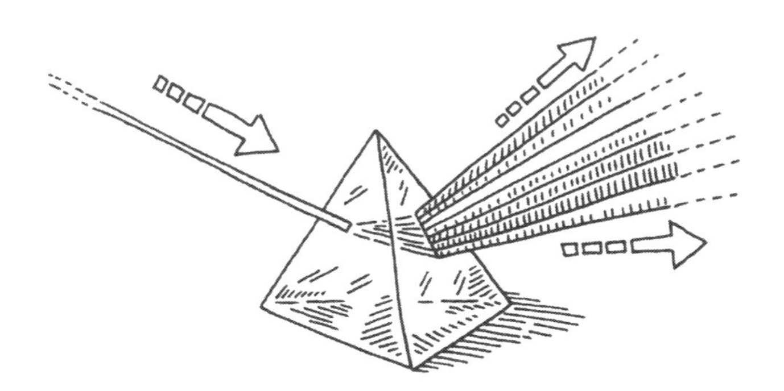

PRISM

by **Amy Ludwig VanDerwater**

White light holds a treasure.
A prism is the key.
I hold it up to sunshine—

Colors!
You are free!
Red and Orange.
Yellow, Green, Blue.
Indigo.
Violet.
Let me look at you!

Bend a rainbow
from white light.
You've been hiding
in plain sight.

THE SHADOW GROWS (AND SHRINKS, AND GROWS)

by **Laura Purdie Salas**

I block sun's rays from reaching ground
as morning floods the eastern sky.
To my west—and sidewalk-bound—
I cast a soundless shadow spy.

At noon, my shadow's incomplete.
The sun above wants it to be
barely bigger than my feet:
a short and stumpy shadow me.

In afternoon, the sun goes low.
I watch my echo dance and spin.
My hazy person starts to grow—
I get to know my shadow twin.

At dusk, as far east as I see,
my legs begin to deepen, stretch.
I reach out for infinity,
my weedy, reedy shadow sketch.

BIG SUN

by **Douglas Florian**

The sun seems BIG,
The BIGGEST star,
But that's because
It's near, not far.
And farther stars,
Though large in size,
Seem just a whisper
To our eyes.

WHAT I KNOW ABOUT THE SUN

by **Eileen Spinelli**

I know that the sun is a dazzling star
far, far from earth. Millions of miles far.
I know that plants, animals, and people
need the sun to grow.
I know that it takes eight minutes or so
for the light of the sun to reach earth.
And that the sun is always shining somewhere
even when it's dark in my back yard.
I also know how the sun shimmers on the pond
where my grandpa takes me fishing.
And how quickly it bakes mud pies on an August day.
I know how the sun brightens everything—even hearts.
And how poets like to sing about it.

SUN-KISSED

by **Guadalupe Garcia McCall**

It is not just your skin that
feels its warm, loving rays.
The sun loved the dirt
who hugged the sun-kissed seed
who unfurled its sun-kissed leaves
and raised the sun-kissed stalk
who produced the sun-kissed wheat
which became a sun-kissed roll
that was baked in a sun-kissed kitchen
and was slathered with sun-kissed butter
because you loved it too.

BESADO POR EL SOL

por **Guadalupe Garcia McCall**

No es solo tu piel la que
siente sus rayos cariñosos y cálidos.
El sol amó a la tierra
que abrazó a la semilla besada por el sol
que desdobló sus hojas besadas por el sol
y cultivó el tallo besado por el sol
produjo el trigo besado por el sol
que se convirtió en panecillo besado por el sol
que fue horneado en una cocina besada por el sol
y fue untado con mantequilla besada por el sol
porque también tú lo amabas.

I LIKE THAT NIGHT FOLLOWS DAY

by **April Halprin Wayland**

What if there was never night—
if it was always light . . . and light?

No dark, no yawn, no closing eyes.
No moon or stars in any skies.

No quiet that the nighttime brings.
I'm sort of scared to think those things.

A sky that's dark as my dog's nose
is just the way things ought to go.

QUEEN OF NIGHT

by **Terry Webb Harshman**

I am the moon, Queen of Night,
riddle wrapped in borrowed light,

a silver spool where dreams unwind,
ancient orb as old as time.

I masquerade; I wax and wane . . .
forever changing yet the same;

I stir the tides with unseen hands;
they ebb and flow from sea to sand.

Father Sun may keep the day;
I ride along the Milky Way . . .

holding court with owls and bats,
moles and voles and backstreet cats.

Within my tent the weary rest;
puppies doze and sparrows nest.

Children dream beneath my light . . .
I am the moon, Queen of Night.

LUNAR ECLIPSE
by **Avis Harley**

When Earth is between Moon and Sun,
it's time for a little shadow-fun.

In a game of heavenly hide-and-seek,
Sun must never be allowed to peek

at Sister Moon—who's kept well hidden
behind Earth's girth. Light's forbidden!

Earth briefly has the power to place
total dark across Moon's face.

Moon finds this game goes to her head;
she blushes a coppery orange-red.

Then Earth will slide her shade away
and let Sun's rays come into play.

LUNAR ECLIPSE
by **Bobbi Katz**

Like a dark old penny
being placed
on top of a bright new one,
Earth's shadow slowly slips
over the copper moon
hiding it completely.
Then slowly,
slowly,
it slides off
as if pushed
by an invisible
finger.

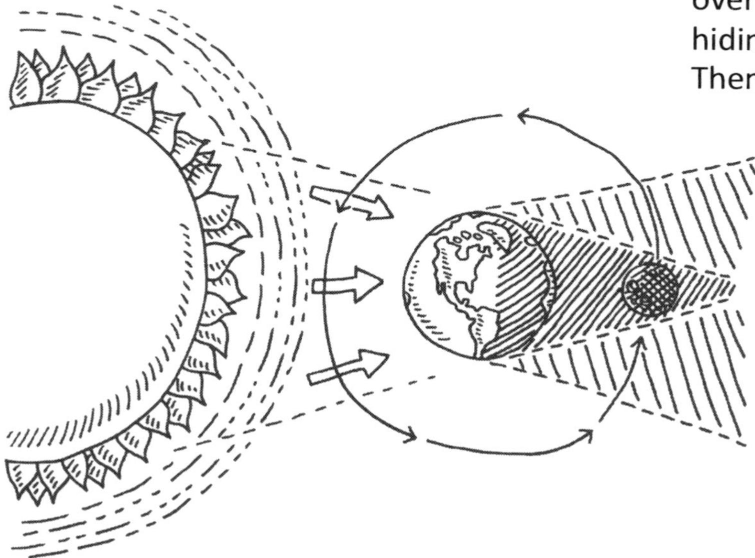

CONSIDERING COPERNICUS

by **Bobbi Katz**

Copernicus, a sage of old,
did not accept what he was told.
He said, "Earth moves around the sun,"
which seemed absurd to everyone,
who saw just how such things were done:
The earth is circled by the sun!
Astronomers and other folk
thought Nicholas was, at best, a joke!
"Sun rises in the east by dawn,
plows through the sky and then is gone
beyond horizons in the west.
Then night
arrives
and people rest.
The sun keeps going while we're sleeping,
rising when the birds start peeping."

The process was so obvious
to ALL—
except . . .
Copernicus.

EARTH'S TILT

by **Douglas Florian**

Earth's axis is tilted—
Its orbit ecliptic.
Exactly why
Remains somewhat cryptic.
Astronomers ponder
A whole host of reasons.
One thing is for sure—
The tilt causes seasons.

GALILEO GALILEI

by **Renée M. LaTulippe**

The stars tell stories
of Galileo Galilei.

A genius with a telescope,
he turned his lens upward
and magnified the moon.

He magnified the moon!
And the planets—
the rings of Saturn
the phases of Venus
the four bright moons of Jupiter.

And craters
and moon mountains
and billions of Milky Way stars.

Billions of stars!
Like Copernicus, he knew
that we spin around
a stationary sun.

Centuries ago,
he turned his lens upward
and magnified a universe
of knowledge.

LOOKING AT THE SKY TONIGHT

by **Janet Wong**

Dad and I look up
at the sky.
He points to dots
and asks what they are.
I say "well . . . stars,"
but he wants
something else,
some different words.
He's tracing a shape
with his finger in space—
and now I see!
It's a measuring cup
or a powder drink scoop—
or something bigger—
smaller than a pot—
but not by a lot—
how about we call it
The Big Dipper?

COMET HUNTER

by **Holly Thompson**

ever since he was a schoolboy
studying the Great Comet Ikeya-Seki
Yuji Hyakutake had wanted
to discover a comet
so he moved to Kagoshima
with clear mountain air
and night after night
while his wife and sons slept
he aimed his huge binoculars
at one patch of eastern sky

memorizing the stars
their exact locations
sketching positions
of constellations
hoping for a rare
blaze of ice and dust

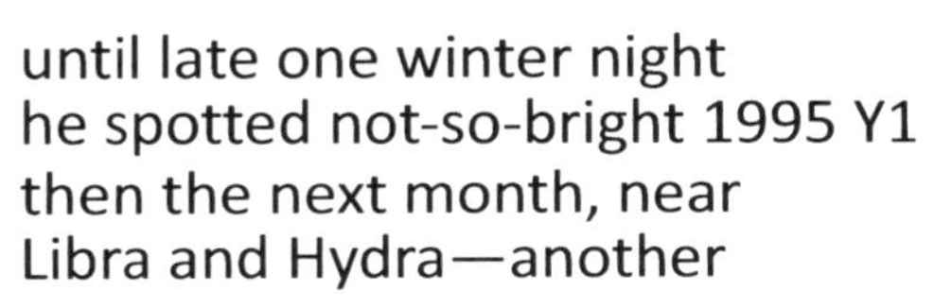

until late one winter night
he spotted not-so-bright 1995 Y1
then the next month, near
Libra and Hydra—another

"I must be dreaming," he thought
as he sketched and photographed
measured and documented
till day dawned on the mountain

"Possible comet," he reported
then waited as top astronomers
at the National Observatory in Tokyo
checked and finally confirmed
he was the first to discover 1996 B2

soon, as the orbit neared earth,
everyone was gazing up
at that glow of ice and dust
with the long sweep of tail—
at the newly named, the truly Great
Comet Hyakutake

ORION NEBULA
by **Mary Lee Hahn**

It's huge.
It's far.
The birthplace of
stars.

It's dust.
It's gas.
Gigantic in
mass.

THE NEO HUNTERS
by **Juanita Havill**

We are hunters
who watch the sky
with telescopes
and camera eye.

We watch for signs
of NEOs
and seek to know
where each one goes.

A Near Earth Object,
an asteroid,
as big as a car,
in the cosmic void
is a rock in space
we want to avoid.

UH OH, PLUTO
by **Jeannine Atkins**

Once Pluto was proud to be called one of nine planets.
But astronomers decided he was too small,
too far from the sun, made unpredictable orbits.
They tore pictures of poor Pluto off walls
and museum halls showed only eight planets.
Happily, Pluto found new friends, whirling balls
of rocks, dust, and ice called comets.
Orbiting whimsically together, Pluto is greatest of all!

Note: Once considered the ninth planet, Pluto was reclassified as a "dwarf planet" in 2006.

OLD WATER

by **April Halprin Wayland**

I am having a soak in the tub.
Mom is giving my neck a strong scrub.

Water sloshes against the sides.
H_2O's seeping into my eyes.

The wet stuff running down my face?
She says it came from outer space!

The water washing between my toes
was born a billion years ago.

LIFE CYCLE

by **Charles Ghigna**

The stream
becomes
the river
becomes
the root
becomes
the tree
becomes
the sky
becomes
the cloud
becomes
the rain
becomes
the stream

WATER ROUND

by **Leslie Bulion**

Raindrops fall
Splink
Splash
Onto soil
Drip
Seep
Into creeks
Trickle
Gurgle
Joining rivers
Whoosh
Flow
To the ocean
Go
Drops
Go!

Evaporate
Waft
Rise
Cool to droplets
Cloudy
Skies
Clouds grow heavier
And then?
Raindrops fall
Begin again!

OCEAN ENGINE

by **Leslie Bulion**

Sun beats,
Ocean heats,
Evaporates
where trade winds meet.
Clouds form,
Tropics storm
Heat's released
Air rises, warm.
Wind belts blow,
Currents flow,
Nutrients mix
So plankton grow,
Spinning on
This Earth we know.

But greenhouse gases slash the time it
Takes to change Earth's fragile climate.

Sun beats,
Sea *over*heats,
Polar ice
Shrinks and retreats.
Shifting rains,
Hurricanes,
Rising seas
Swamp coastal plains.
Winds squall,
Currents brawl,
Biodiversity
Numbers fall—
Which engine's running?
It's your call!

WATER ENGINEERED

by **Sara Holbrook**

Water is pumped all around.

Engineers plan
how it's captured,
channeled,
and hosed
through huge valves,
        (open and closed)
    designed to make water rush
when I need a shower,
    a drink,
        or a flush.

Water is pumped all around.

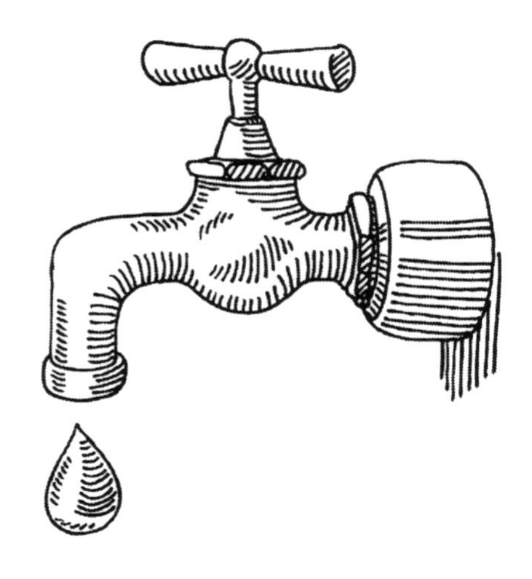

WATER

by **Kate Coombs**

Turn on the tap
and the water flows.
Does anyone know
where the water goes?

Turn on the tap
and the water comes.
Does anyone know
where the water's from?

Water is clean
and water is cool,
living in rivers
and raining in pools.

Yet water can trickle
and wells can dry up
till there's nothing left
to fill the cup.

Today there is water
when we turn the tap on.
But what will we do
when the water is gone?

OH WATER, MY FRIEND

by **Guadalupe Garcia McCall**

Are you scared?
Is that why you run
to rivers, to streams, to lakes?
Do you feel safer pooled in ponds?
Does it hurt to boil?
Does your anger roll and roil?
Is that why you recoil from the heat and sun?
Do you feel trapped, contained, restrained?
Is that why you weep and seep
through window panes?
Does it thrill you, fulfill you,
to dissipate—evaporate?
Does it feel weird to leave the earth,
to rise above the rest?
Do you get dizzy hovering in the heavens?
Are you afraid to fail—to fall?
Aren't we all.
Aren't we all.

AY AGUA, MI AMIGA

por **Guadalupe Garcia McCall**

¿Tienes miedo?
¿Es por eso que corres
hacia ríos, arroyos, lagos?
¿Estás más segura encharcada en estanques?
¿Te duele cuando hierves?
¿Es tu rabia que rueda y se agita?
¿Por eso retrocedes del calor y del sol?
¿Te sientes atrapada, contenida, refrenada?
¿Es por eso que lloras y goteas
sobre el vidrio de las ventanas?
¿Te emociona, te realiza,
poder disiparte, evaporarte?
¿Te sientes rara al dejar la tierra,
al ascender más alto que los demás?
¿Te mareas flotando en los cielos?
¿Tienes miedo de fracasar, de caer?
Nosotros también.
Nosotros también.

OCEAN EXPLORER SYLVIA EARLE

by **Leslie Bulion**

She walks deep down on the ocean floor,
Where no one has ever walked before,
Then jumps in a submarine to explore more.

She teaches us each to do our part,
We *will* protect our oceans. We're smart!
We can save planet Earth's true blue heart.

CLOUDS

by **Kate Coombs**

I saw one little cloud
that looked like a wish,
but now there's a crowd
like a school of white fish.

Clouds can turn red at sunset
or shine with gold light.
Sometimes dark clouds growl
with thunder at night.

There are clouds flat as paper
and clouds fat as geese,
clouds built like staircases,
clouds soft as fleece.

But clouds *should* look wet—
and do you know why?
All clouds are secretly
lakes in the sky.

WEATHER MAP

by **Joan Bransfield Graham**

A crazy quilt of suns and sleet,
weather patterns that repeat,

the meteorologist paints away—
a chance of rain, a sunny day.

Hot and humid, cold, wet, dry—
the weathercaster's practiced eye

checks for clues, consults the chart,
knows the online sites by heart,

plots the future from the past.
How long will that Cold Front last?

Arrows show its bold advance
in the daily Weather Dance.

Clouds and suns are rearranging . . .
weather's face is always changing.

DOG IN A STORM

by **Stephanie Calmenson**

Yo! It's me—Buster the Brave.
Wait. I feel a storm coming.
The air is hot. It's humid.
Winds are blowing.
Clouds are rolling in.
The air is suddenly getting cooler.
KABOOM!
Thunder! Lightning! Rain, rain, rain!
I'm scared. I dive under the bed.
The weather reporter says,
"Thunderstorms may come when
cold air pushes warm air up—"
KABOOM! KABOOM! KABOOM!
Then the sky gets lighter.
The world gets quieter.
The rain stops.
I come out from under the bed.
Yo! It's me again—Buster the Brave.

RAIN GAUGE

by **Anastasia Suen**

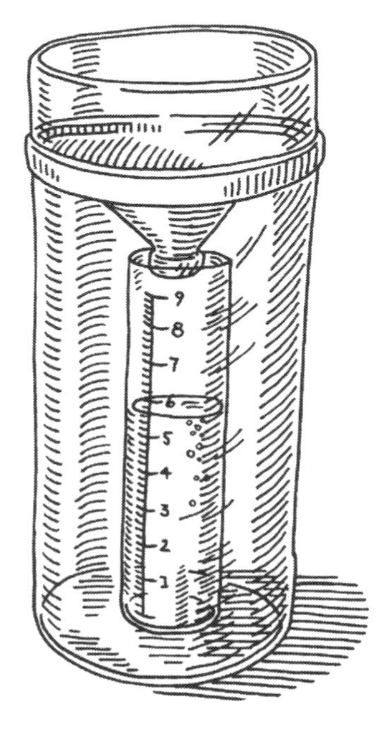

I wonder how much it rained last night?
The rain gauge will show me the number.

I start at the bottom and count my way up.
I count up to the top of the water.

That's all? It rained for hours and hours.
I thought the number would be much higher.

Maybe tomorrow. I turn the gauge
upside down and pour out the water.

I wonder how much it will rain next time?
The rain gauge will show me the number.

THIS WEEK'S WEATHER

by **Janet Wong**

Monday's temperature was 93.
Way too hot.

Tuesday's high was 90.
Better, but not a lot.

Wednesday it cooled down.
It was 75—just right.

And yesterday, a storm came.
It rained cats and dogs all night!

It rained so much
it was like the sky sent down a river.

It was windy and only 50 degrees—
I felt myself shiver.

How could the weather change so much
in a week (or less)?

This is why we have
weather news, I guess!

CLIMATE VERSUS WEATHER

by **Joan Bransfield Graham**

Climate's your personality,
weather is
your mood—
a warm and sunny outlook,
with occasional
attitude.

Low pressure grumbles in with rain,
an atmospheric
pout.
High pressure smiles and saves the day,
sweeps the stormy
out.

Where are you living on this globe—
your latitudinal
position?
Location has a lot to do
with your climatic
disposition.

RIDDLE FOR A DRY DAY

by **Irene Latham**

Sun without rain
day in, day out.
Grass browns, ground frowns.
I am a DROUGHT.

DISASTER RIDDLE IN A HURRY

by **Irene Latham**

Sloping avalanche of earth
hurtling down a mountainside,
I smother homes and choke trees.
I am a sudden LANDSLIDE.

DISASTER RIDDLE UNDER PRESSURE

by **Irene Latham**

When I belch smoke and start to grumble,
Please pardon the rude interruption.
You never know if or when I'll blow –
I am a VOLCANIC ERUPTION.

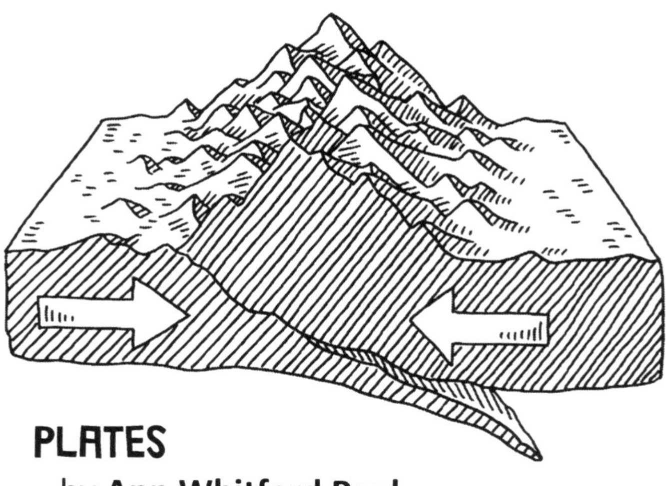

PLATES

by **Ann Whitford Paul**

We live on plates—
not the eating-off-of kind—
great slabs of rock
that slowly move
deep
        deep
              deep beneath our feet.

Sometimes some plates—
not the eating-off-of kind—
collide . . . slide . . .
trembling the very ground we walk—
EARTHQUAKE!

Everything shakes.
Bridges split,
chimneys crack,
towers crumble,
and inside houses,
plates—
the eating-off-of kind—
shatter,
break.

Note: Tectonic plates move in an area called the lithosphere, which includes the earth's upper mantle and crust. This movement is due to pressure and heat from the mantle (convection currents).

HARBOR WAVE AT HILO: TSUNAMI SURVIVOR

by **Carole Boston Weatherford**

Hawaii, April Fools' Day 1946.
Kazu and his classmates
see a strange tide suck the water
from the beach at Laupahoehoe.

Curious children scamper
to the shore. They see a rainbow—
butterfly, parrot and puffer fish, lion fish
and triggerfish—flopping on the sand.

Then, the earth rumbles, the ocean
roars and a monster wave rushes in.
The mighty wave shreds school buildings
and sweeps Kazu and his friends to sea.

The island a speck in his eye,
Kazu paddles wildly, treading water
amidst sea turtles, whitecaps, and splintery
wreckage of boats and buildings.

Tired, Kazu grasps a wooden board; makes it
his raft. Can't sleep for fear of sharks.
One long day later, a sailor saves him.
Kazu's mother calls off the funeral; fixes a feast.

Note: A tsunami is a series of extremely long, traveling ocean waves caused primarily by earthquakes below or near the ocean floor. Tsunami waves can move at more than 500 miles per hour and can reach heights of more than 100 feet in coastal waters. During the 1990s, more than 4,000 people were killed by ten tsunamis. Eighty percent of tsunamis occur in the Pacific Ocean. However, tsunamis also threaten coastlines off the Indian Ocean, Mediterranean Sea, Caribbean region and even the Atlantic Ocean. This story is based on the real-life experience of Yoshikazu Murakami, a Hawaiian teenager who survived being swept to sea by the tsunami of April 1, 1946. Caused by an earthquake in the Aleutian Islands, this tsunami claimed 159 lives and destroyed boats, piers, roads, bridges, railroad tracks and buildings.

RIDDLE FOR A WET DAY

by **Irene Latham**

I overwhelm and overflow
with raging waves and sheets of mud.
I leave behind disease and sludge.
I am an unexpected FLOOD.

HURRICANE HIDEOUT

by **Janet Wong**

The weather reporter
says it might be
a really bad hurricane,
Category 3.
Her words
are jumbling up
in a ball:
storm surge,
flood watch,
winds,
eye wall,
low pressure,
cyclone,
large heat engine.

All I know is
I'm going to get
IN
the safest part
of our house,
the tub.
I'm going in there
with the stuff I love:
radio, pillows,
candy, light, book.
Watch the news
but don't make me look.
What? You filled the tub
to the top
in case our water
gets shut off?

No problem.
I can deal with it.
You'll find me
in the hall closet!

TORNADO!

by **Carole Gerber**

Warm, moist air drifts toward the sky;
gets caught in cold air speeding by.
Vicious, raging rains erupt.
Lightning flashes, wind speeds up
and shapes into a funnel form.
Tornado!
Deadly product of the storm.

MOUNT ST. HELENS, WASHINGTON

(from the point of view of Pocket Gopher)

by **Carmen T. Bernier-Grand**

My underground home is in the shoreline of Spirit Lake.
Near the serene cone-shaped, snow capped Mount St. Helens.

March 24, 1980
Small quakes. Something is welling up inside the mountain.

March 27
A geyser of ash blasts out the mountain peak.
Ash dusts my home entrance. A blue flame. No lava.

May 7
Stronger ash eruptions. Frequent quakes. Steam puffs out.
The mountain is shredding its now blackened snow.

May 12
Earthquake rattles the ground! An avalanche snaps loose.
A chunk of ice almost hits me.

May 18
Large earthquake! A gigantic mushroom cloud rises high.
Day turns into night. Lightning! A blanket of ash. Flying rocks.
I take cover at home. I dig, away from the heat.
Deafening roar of a sea of logs.

May 19
Quiet. I peek out. Is this what the end of the world looks like?
I burrow into the forest's built-in cellar,
Eat roots of plants that have disappeared above ground.
I push old soil from below to the surface.

September 24
I till the ground until branches break through the ash mantle.
Blushing pink fireweed! Purple lupine! Alder trees!
I welcome back Bird, Bear, Bobcat, Elk, Rabbit, and Grasshoppers.

A new lava dome slowly rises in Mount St. Helen's crater.
The mountain is rebuilding itself. I till to make it pretty again.

STORY ROCKS
by **Susan Marie Swanson**

There are stories in the wind,
in rushing rivers,
even in the dirt on our shoes.
We ask questions.
We make observations, chart patterns,
and tell Earth's stories the best we can.

Geologists tell of Earth long ago.
(" Once upon a billion years ago, there was a rift
in North American continent . . .")
Gauges and switches glow
on instruments in a geology lab.
Here is a researcher working at his computer
after dark, analyzing data
and writing one part of one story
he is trying to tell.
His face shines in the light of the screen.

More pieces of this billion-year-old story
can be found in rocks on Lake Superior's shore.
That's why this scientist
has maps and hiking boots waiting by his door.
He has checked his rain gear,
repaired the rock drill,
exchanged emails with his research team,
and packed candy bars for everyone.
They'll unwrap them next week,
when they're resting on story rocks
at the edge of the big lake.

GLACIER
by **Kate Coombs**

The great ice serpent
dragged its cold coils,
scraping its tail
against rocks,
leaving its sign behind,
a curving valley trail.

GEOLOGIST

by **Betsy Franco**

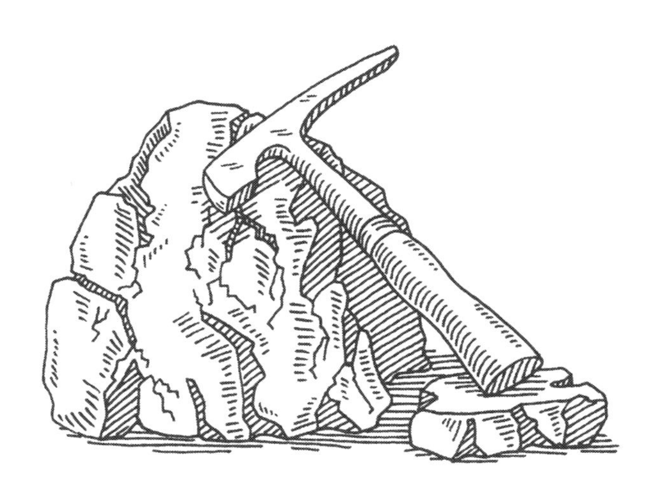

My mother and I are
out in the field again
at Medicine Lake Volcano
where bald eagles
own the sky.

Adding hammer sounds
to the semi-silence,
we break rocks—
some filled with white flecks
some with yellow—
to unlock the story of
the lava's flow.

Walking up a cinder cone
our feet sink into
fine sand.
Below us lies
the dry lava bed
we map in
four dimensions:
width, length, depth,
and time.

In a soft voice
my mother says
she knows
this place better
than anyone.

MY ROCK

by **Ken Slesarik**

My rock is cold,
gray, white, hard,
small, rough
and round.
Are rocks living?

SEASHELLS IN THE SKY

by **Laura Purdie Salas**

Mt. Everest reaches five miles high
A layer of seashells trapped inside
How'd they get from sea to sky?
When mountains formed, they hitched a ride

Sixty-five million years ago
Creatures filled a shallow sea
They drifted in the water's flow
Then crumpled into shell debris

A broken layer on ocean's floor
With time, they formed the limestone rock
Earth pushed Earth—a shoving war
And mountains rose in massive block

So climb that icy mountain peak
That world where nothing living dwells
Dig through frozen meadows bleak
And bring a bucket—for the shells!

Note: A layer of limestone, formed from broken seashells, was once at the bottom of the sea. As pieces of the Earth's crust pushed against each other, that layer of limestone was part of the crust that buckled and rose and formed the Himalayan Mountains. The Yellow Band is the most famous stripe of limestone in Everest.

WHAT IS A FOSSIL?

by **Rebecca Kai Dotlich**

A picture,
a puzzle,
a clue that is known
of prehistoric days;
a riddle in stone.

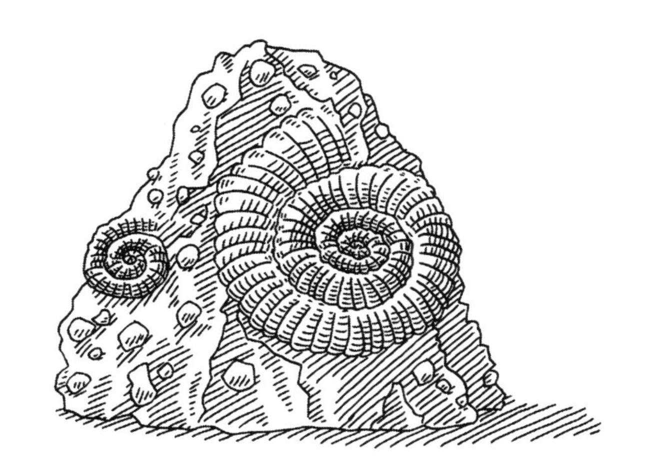

STONE, SEA, AND SILENCE

by **Jeannine Atkins**

As birds swept color and song through
Appalachian Mountain sky, a girl wandered,
found a fossilized shell and wondered
how it got so far from the sea.
How does a shell turn to stone?

Wonder leads to science. Rachel Carson learned
how stone and soil tell stories without words.
She studied the shifting sea, which one spring day
touched shores where birds had stopped singing.
Such strange stillness tells a story, too.

Rachel listened, looked, and studied.
She wrote that people must change so land
and sea can change in ways that keeps life safe.
Science leads to courage, which circles
back to wonder, and tells a story, too.

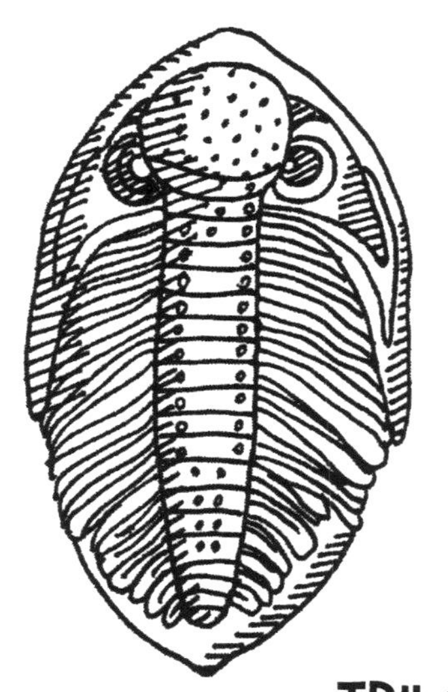

Note: Rachel Carson (1907—1964) was a biologist best known for her book, *Silent Spring,* which showed how DDT, a chemical used to kill insects, also poisoned the birds who ate the insects, and endangered all animals, including humans. DDT was made illegal largely due to her warnings.

TRILOBITE

by **Mary Ann Hoberman**

Tri-lo-bite, tri-lo-bite, that is my name
My body's three-sectioned; my name is the same.
I lived when the seas covered mountains and plains.
Now I am gone but my fossil remains.

SOIL INVENTORY

by **Kate Coombs**

Fifteen dissolving
veins of leaves,

fading roots,
disappearing shoots,

six wing cases, thoraxes,
legs thinner than stems,

the night earth of eleven
bugs, beasts, and birds,

a scrap of paper slowly
returning to wood,

the toe bone of a mouse,
a dried bit of worm,

tiny variegated rocks
that were boulders once,

a fragment of snail shell,
two marigold seeds,

and from a lost bracelet,
one red plastic bead.

MAGIC SHOW

by **Juanita Havill**

I don't use a wand.
I don't use a hat,
nor bright silk scarves.
All I need is a vat.

Into my vat—
with slits on the sides—
go corn husks, dead plants,
leaves that have dried,

tea bags, coffee grounds,
veggie peels galore,
old straw, sawdust,
egg shells, and more.

Water if needed.
Be sure to add dirt.
Turn with a pitchfork.
Better than dessert.

No bunnies or doves
in this magic show.
Abracadabra!
Let the compost flow.

SHADE-GROWN

by **Margarita Engle**

One village away: a deforestation disaster.

But here there is no slash-and-burn.
No clear-cuts.
Beneath towering rain forest trees,
farmers plant cacao.

Toucans and parrots weave nests.
Monkeys and marmosets play.
Orchids bloom.
Butterflies flutter.

A plentiful harvest of cocoa pods
will be roasted to make sweet chocolate.
An eco-agricultural success story.
Delicious!

DEAR RACHEL CARSON

by **Mary Lee Hahn**

Dear Rachel Carson,

We went to the organic farm yesterday
and learned about you, and why they don't spray
chemicals to kill bugs that eat up their crops:
the balance in nature goes from bottom to top.

You warned in your book, *Silent Spring*, long ago,
humans must always be sure that they know
what the impact will be on *all* living things
when we do things to benefit us, human beings.

Someday when we're scientists, we'll think of you
and remember your teaching in all that we do.
In our work to help humans we'll never forget
that we're only one part of the life on our planet.

Yours truly,
Miss Smith's 1st Grade Class

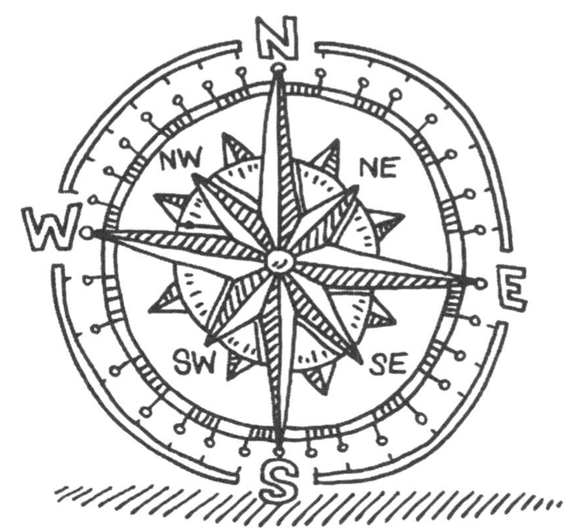

SHEN KUO
(1031—1095)

by **Janet Wong**

Almost a thousand years ago
a Chinese scientist named Shen Kuo,
geologist-cartographer-astronomer-engineer,
discovered fossilized shells
hundreds of miles inland
that made it clear the shoreline had moved.
Petrified bamboo convinced him
that climate change was happening.
But people did not want to hear these things.
Instead he became known for the idea
that true north is not magnetic north.
His magnetic needle compass was worth
spices, gold, jewels—even a giraffe—
as explorers later sailed to Africa and back.

Climate change and the shifting sea:
who would choose such mundane news
over promises of spices, gold, and jewels?

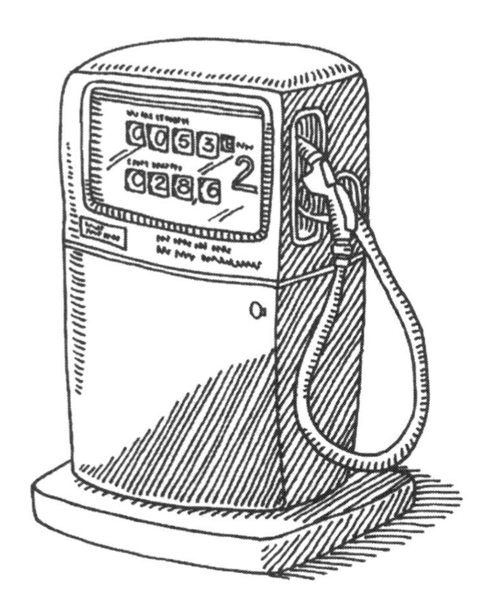

FOSSIL FUELS

by **Janet Wong**

They're talking
about fossil fuels
on the news.
I ask Pop
what those are
and he says
fossil fuels
are oil and gas and coal
made from plants and plankton
that sank down in the water
and got covered and cooked
in a thick mud crust
for millions and millions
of years.

No wonder
it costs so much
to fill up our car:
our fuel
took millions and millions
of years
to make!

AUNTIE V'S HYBRID CAR

by **Janet Wong**

Auntie V drives a hybrid car.
It has a little screen
that shows green leaves
sprouting on a tree
when she's driving electric.
If she goes too fast,
she uses gas,
and leaves fall off.
Go slow, Auntie V—
and grow a tree for me!

WHAT MAKES A TURBINE TURN

by **Steven Withrow**

The formless force
that waggles a flag
and shapes a ghost
from a plastic bag

and levitates
a dragon kite
and wrestles with
the trees at night

is named the same
as that airy motion
which blusters
over field and ocean

and charges up
electric motors
with each revolving
round of rotors.

When next you see
a three-armed beast
who might be facing
north-northeast

don't worry if
you feel thin-skinned.
"It's just my pinwheel,"
says the wind.

Note: Wind is not the only thing that makes a turbine turn; there are also solar turbines, gas turbines, steam turbines, and more.

SOLAR POWER

by **Susan Blackaby**

Solar power! Feel the heat!
Light the lights along the street,
run the engines, fuel the cars,
turn the turbines with a star!

Quick! Let's build a head of steam,
fire up some clean machines,
set in motion cranks and cams,
swirling gears and pumping dams.

Solar cells turn light to juice—
electron transfer on the loose!
Tap this energy in space!
The sun can win Earth's resource race.

RESOURCES RULE!
by **Susan Blackaby**

As stewards of the biosphere,
it is our job while we are here
to nurture and rejuvenate
the resources it takes to make
the things we use to meet our needs,
our comforts and necessities.

Ore, minerals, and oil that hide
in veins and caverns deep inside
Earth's firm and fragile outer crust
are quarried, mined, and drilled by us.
These raw reserves may disappear
without responsibility and care.

Preserve the precious watershed—
the run-off and the river bed.
Wise use in cities, farms, and towns
keeps water pure upstream and down,
so waterways run clear and clean
along their course from spring to sea.

Conserve the forests, lush and green,
from humus up to canopy.
Protect the woodland denizens
that need the trees for nests and dens.
Every time you hike around it,
leave Earth better than you found it.

The rich resources on our planet
can be things we take for granted,
but they'll be in short supply
if globally we don't comply
with simple rules we all respect:
reserve, preserve, conserve, protect.

RECYCLING
by **Susan Blackaby**

Collect the daily scraps and clippings,
gather up the bits and snippings:
Paper, plastic, glass, and tin—
all of these go in the bin.
Once it's sorted and inspected,
so-called waste is redirected.
Think of all the things that you
can make from useful stuff you threw
away!

WE NEED GREEN SEAWEED!

by **Margarita Engle**

The carbon dioxide/oxygen cycle
is just a circle. Pretty simple.
My nose needs Os,
and green plants crave Cs,
so we take turns recycling
the same airy breeze.

It's easy, as long as no one
makes the ocean dirty, or chops down
all the trees.

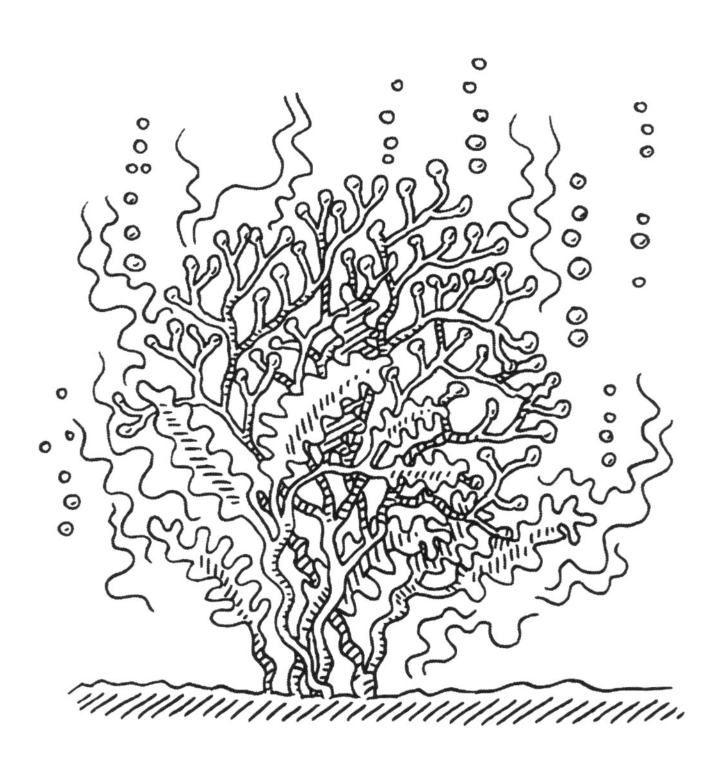

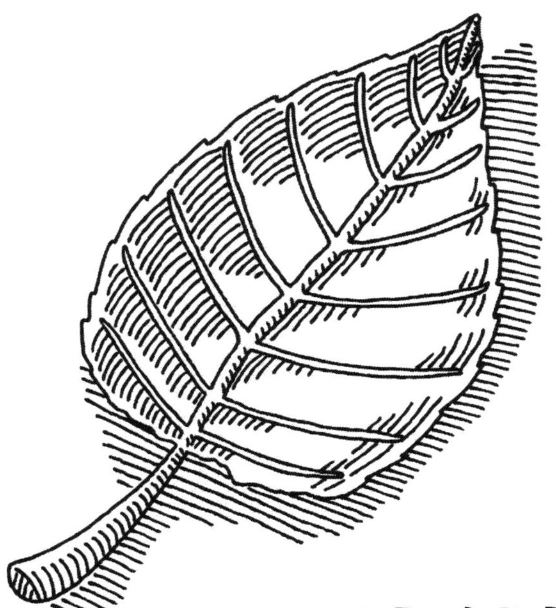

PHOTOSYNTHESIS

by **Marilyn Singer**

You start with lots and lots of air
(preferably clean),
the right amount of water,
and, oh yeah, you must be green.
But it won't work, you'll make no fuel,
try with all your might,
if you're not some kind of a plant,
and you haven't any light.

COOL FOOD FOR THOUGHT

by **Sara Holbrook**

Plants! The original solar panels,
whether swaying or standing still
transfer
blue and red wavelengths of sun
into 30 shades of green
known as chlorophyll.
Whether you pluck your food from a tree
or eat it on a bun,
of all the lion-human-chicken
links in the food chain,
plants are number one.

But plants not only feed our stomachs,
they also scrub the air,
converting carbon emissions
into the oxygen we share.
Sustaining plants are an army of organisms,
7 billion in every teaspoon of healthy soil,
together they feed us and cool the atmosphere
so we won't starve
or start to boil.

YOUNG AND OLD TOGETHER
by **Margarita Engle**

I love to help Grandpa in his garden,
planting tiny radish seeds
so we can watch the swift growth
of leaves and stems,
like green towers
on top of
tasty
red
roots.

JÓVENES Y VIEJOS JUNTOS
por **Margarita Engle**

Me encanta ayudar a mi abuelo en su jardín,
sembrando semillitas de rábano
para mirar cómo crecen tan rápido
las hojas y los tallos,
como torres verdes
encima de
sabrosas
raíces
rojas.

WHAT WE EAT
by **Joseph Bruchac**

It has been said
we are what we eat.

Perhaps that is so
but what I think now
is that in the end
it all comes down
to the simple words
my Indian grandfather
spoke one late summer
as we knelt
in the garden
to pull up
sun-gold carrots.

Brushing them
on his sleeve,
Grampa said to me
"We all got to eat
a ton of dirt"
his words taking me
back to the soil
that feeds us,
that we feed.

ACCIDENTALLY ON PURPOSE

by **Linda Sue Park**

A hundred thousand lifetimes ago,
somebody figured it out:

If you planted the corn in the middle,
and the beans around it, the bean vines
could climb up the cornstalk,

which meant less work because
you didn't need to put in poles
to support the beans, and hey,

it turns out that the corn takes nitrogen
out of the soil, but the beans put it back,
so the soil stays healthier.

Most of the time, we don't know
what we don't know. It's only later
that we realize it. *head slap*

They tasted good together, beans and corn.
Succotash. Tortillas with refried beans.
Baked beans and cornbread,

and surprise, surprise: each by itself
is missing something, but together
they make a complete protein.

head slap again.

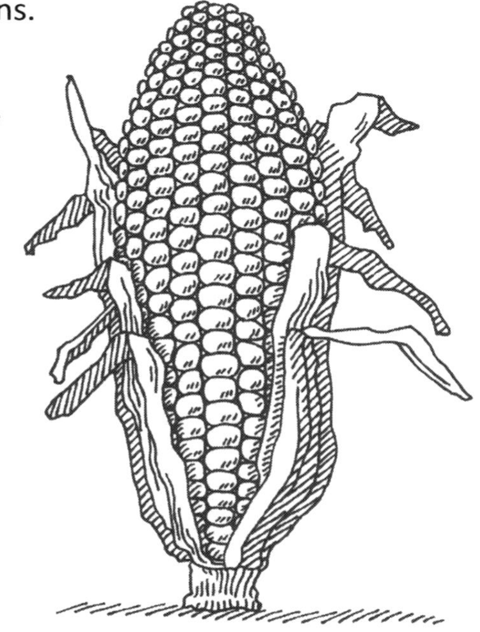

BUTTERFLY GARDEN

by **Jane Yolen**

The butterfly sips—
though not through lips—
for hour after hour.
It finds and takes
its nectar shakes
from every kind of flower.

POLLINATION

by **Margarita Engle**

Bees
need flowers
and flowers
need bees.

We need sweet fruit
and crunchy nuts
to eat . . .

so we need
our friends

the helpful
bees.

POLINIZACIÓN

por **Margarita Engle**

Las abejas
necesitan a las flores
y las flores
necesitan a las abejas.

Nosotros necesitamos frutas dulces
y nueces crocantes
para comer . . .

por eso necesitamos
a nuestras
útiles amigas
las abejas.

A BIOLOGICAL COMMUNITY/UNA COMUNIDAD BIÓLOGICA

by/por **Margarita Engle**

Students trade treats at lunchtime.
Los estudiantes intercambian bocaditos.

Ants on the soccer field discover cookie crumbs.
Las hormigas descubren en el campo de fútbol migajas de galletas.

Sparrows under the tables find plenty of food.
Los gorrioncitos encuentran mucha comida debajo de las mesas.

Flies in the trash cans eat what no one else wants.
Las moscas de los basureros comen lo que nadie quiere.

Every organism has its own niche.
Cada organismo tiene su lugar propio.

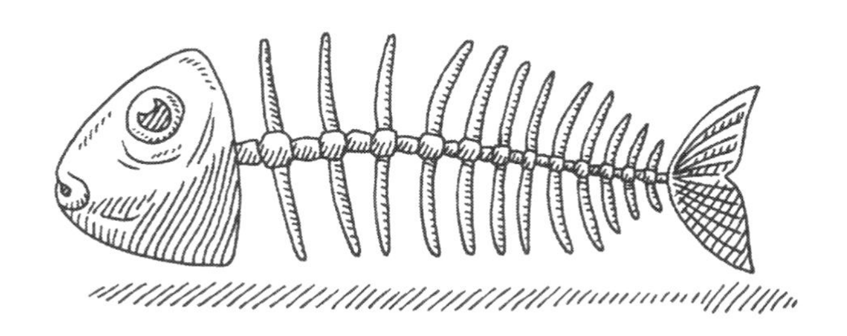

SNACK

by **Jane Yolen**

Here is a grackle.
Here is a snail.
Here is the way
that a bird cannot fail
to ease its sharp hunger.
This quick little snack'll
be one little snail
for a rather large grackle.

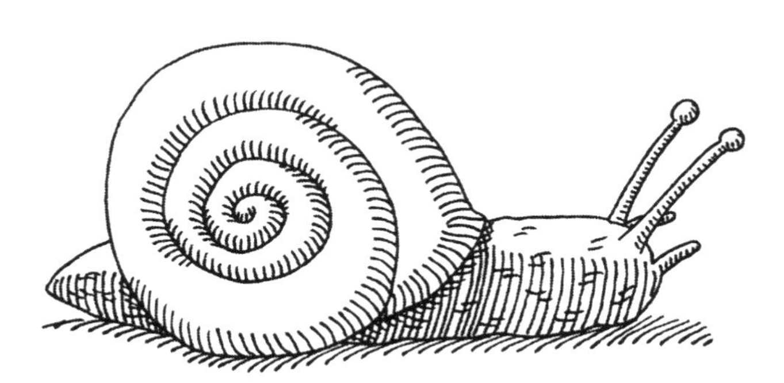

ALLIGATOR WITH FISH

by **Jane Yolen**

So many fish,
so many teeth,
dinner is always
just within reach.
Like a great colander,
like a big sieve,
the gator takes all
that the river will give.
He opens his mouth
in a big toothy smile.
Fish disappear.
He'll be full—
  for a while.

SEVEN WORDS ABOUT AN ALLIGATOR

by **Jane Yolen**

Silently floating,
Silently gloating,
Not a log.

TIDE POOL

by **Jane Yolen**

Between two rocks,
A world made small,
The naturalist
Can't count it all.

The population
Does sustain
Itself with neither
Milk nor grain

But pulls its living
From the sea
And from the sun
So naturally

That whelk and slug
And sponge and prawn
And crab and star
Live on and on

Between the rocks,
Between the tides,
Where life so bright
Precariously rides.

HAPPINESS IN THE DESERT

by **Joy Acey**

is 100 year old saguaros,
cholla, prickly pear cactus
and Palo Verde trees.

It's cactus wrens, gila woodpeckers,
quail, bats and
road runners.

It's scorpions, tarantulas,
gila monsters, coyotes
and javelinas.

It's monsoon thunderstorms
and rainbows
after the rains.

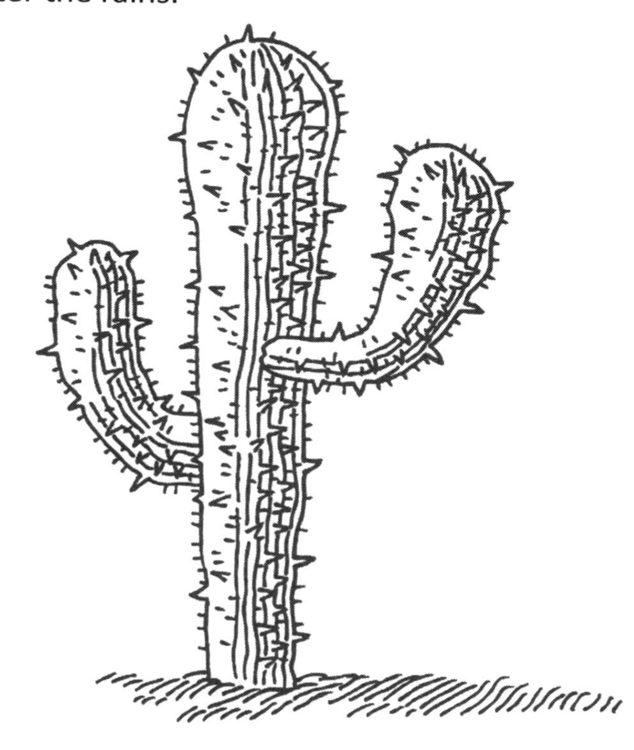

THE RAIN FOREST

by **Bobbi Katz**

It's a weaving—growing, breathing.
Huge trees form a canopy.
Within this leafy rooftop,
there's more life than we can see.
Above, a few trees poke through,
where Harpy Eagles look for prey.
The rain forest restaurant
serves them banquets night and day.

In the pulsing understory
high above the forest floor,
there are striders, swingers, gliders—
jaguars, monkeys, bats, and more.
There are countless birds and insects,
we cannot name them all.
Fruits and flowers, butterflies—
flying,
        crawling,
chewing,
        calling—
clicks and whispers—
screeching cries.

TROPICAL RAIN FOREST SKY PONDS

by **Margarita Engle**

No space is ever wasted.
Each species must find
its own niche.

At the top of a swaying tree, air plants cling
to high branches, seeking sunlight.
Their dangling roots absorb moisture
from drifting mist, instead of soil.
At the base of each leaf of an air plant,
inside a small puddle of rain,
tadpoles turn into frogs,
insects swim,
and little crabs clack
tiny claws.

No space is ever wasted
in this forest
of surprises!

SQUIGGLES

by **Michael Salinger**

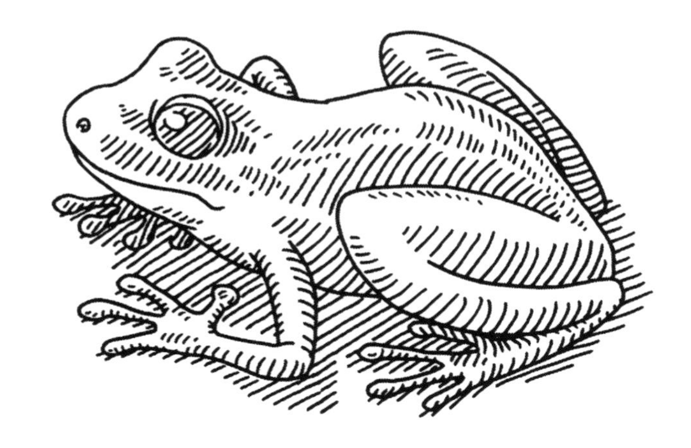

I caught some squiggles in the pond
And put them in a big jar
I gave them bits of earthworms to eat
'Cause they looked kinda starved
They began to grow real fat
And as their bodies spread
Legs popped out of their sides
And eyes bulged from their heads
Their squiggle tails disappeared
They were no longer polliwogs
My squiggles are all gone
Now what can I do with these frogs?

Note: Remember to release your frogs
back to the wild, where they were collected.

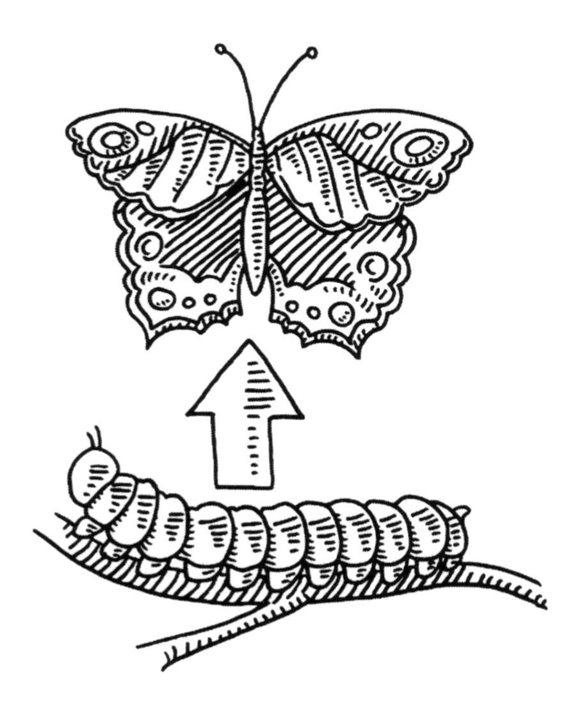

BECOMING BUTTERFLIES

by **Jeannine Atkins**

One tiny egg breaks.
A caterpillar comes out,
eats her former shell,
and sheds her skin.

The caterpillar eats more,
grows more. Her skin splits again,
before she spins
a safe, still chrysalis.

At last, a butterfly breaks through.
She unfolds wet wings,
flutters, seeks nectar,
flits, catches another's notice.

Finally (or is it first?), she finds
a leaf where she lays new eggs,
small as pencil points.
. . . One tiny egg breaks.

CiCADA
by **Guadalupe Garcia McCall**

I dug my toes into the dirt
and felt a tickle at my heel.
A pretty nymph had crept up there
and made me jump and squeal.
"What do you want from me?"
I asked the young cicada.
"I need to climb up on that tree,
to feel the sun, to have some fun,
to shed my skin, and rest and sing.
I need some time to think,
to shake the daze from all that sleep,
and dust off my brand new wings."
So I let her climb upon my finger
and placed her gently on the bark
that she may crawl up to the top
and someday soon enjoy the park.

CHICHARRA
por **Guadalupe Garcia McCall**

Metí los dedos del pie en la tierra
y sentí cosquillas en el talón.
Una ninfa bella se había trepado allí
y me hizo brincar y gritar.
—¿Qué quieres tú de mí?—
le pregunte a la joven chicharra.
—Necesito subirme a ese árbol,
sentir el sol, disfrutar,
mudar de piel, descansar y cantar.
Necesito un poco tiempo para pensar,
sacudir el aturdimiento de tanto sueño
y desempolvar mis alas nuevecitas.—
Por eso la dejé treparse a mi dedo
y la puse sobre la corteza cuidadosamente
para que gateara hasta la copa del árbol
y algún día en breve gozara del parque.

CiCADA MAGiC
by **Heidi Mordhorst**

First, the news:
cicadas are coming!

Then, the holes:
one, three,
fivethirteenseventeenfifty-oneseventy-three
hundreds of little eruptions in the mud.

Next, the shells:
brittle, hollow, yellow-brown;
perfect casts, fresh fossils:
blessedly motionless.

Not for long:
the first ones appear
with Martian-red eyes
at face-height on the front porch.
They litter the pavement,
scatter the windshield.
The lawn crunches underfoot.

And yet:
they are friendly,
allowing us to catch, keep,
compare, even wear them.

And now, the song:
subtle thrumming at morning,
slowly swelling to a throb that meets
the beating of the sun all afternoon.

CAMOUFLAGE

by **Margarita Engle**

When swift eagles strike from above,
and fast jaguars leap from below,
the gentle sloth stays alive by being
SOOOO very SLOOOOW
that algae and moss turn his fur as green
as a leafy branch
with no place to go.

No wonder he always hangs upside-down
with a sly grin, like a tricky clown.

NO HURRY

by **Linda Ashman**

In terms of speed,
this sluggish breed
will never take the crown.

But wins first prize,
endurance-wise,
for time spent upside down.

Note: The sloth is the world's slowest mammal. It spends most of its time hanging in the boughs of trees in South America, and moves very slowly and awkwardly when on the ground—somewhere between 5 to 15 feet per minute.

SNAKE TRAITS

by **Linda Ashman**

The slender,
hissing cobra
is quite famous for its bite.

But the wily anaconda
would much rather
squeeze you tight.

Note: The king cobra, found in Southern Asia, is the longest venomous snake, growing up to 18 feet. If threatened, it spreads its hooded head and raises the front of its thin body to stand 5 feet tall. Although impressive, the cobra seems rather small compared to the anaconda of South America, which is the world's heaviest snake. This constrictor grows more than 30 feet long and can weigh close to 550 pounds.

THE LEOPARD CANNOT CHANGE HIS SPOTS

by **Lesléa Newman**

The leopard cannot change his spots
Into stars or polka-dots.
The tiger cannot change her stripes
("I wish I could," she sometimes gripes.)
The scales remain upon the fish
Though that is sometimes not his wish.
The kangaroo can try and try
But she will never learn to fly.
The cat can't bark, the dog can't purr,
The rabbit cannot change her fur.
The frog can leap but he can't walk,
The lark can sing but she can't talk.
Since we can't be what we are not,
Let's all be grateful for our lot.

THE LION AND THE HOUSE CAT

by **Mary Lee Hahn**

different strength
different size
same chin
same eyes

different mane
different stride
same stretch
same pride

GRAFTING

by **Janet Wong**

We have an old red apple tree
here in the yard of this new house
but the apples don't taste tart enough.
(We like them tart for pie.)
Dad says it's an easy fix.
Today we drove to our old house
and asked to cut some branches from
the apple tree we always loved.
They look like plain bare sticks to me
but Dad says they're just what we need.
We'll graft them onto our tree here,
cut and line the pieces up,
tie them tight and seal with wax,
keep it moist, and hope and hope.
If the grafts grow fine, come harvest time,
we'll have our tart apple pie!

INHERIT TENSE

by **Charles Ghigna**

My family tree is rooted
Deep within my skin;
My eyes look like my mama's,
I have my daddy's chin.

I draw just like my grandma,
I sing like Uncle Lee.
The only question now remains—
What is left of me?

HAND-ME-DOWNS

by **George Ella Lyon**

I look like my mother.
My brother looks like my dad.
Mom looks like her mother.
Dad looks like Grandpa Tad.

Mom says this is genetic
but I don't know what that means.
She says that I'll learn someday
when I study genes.

I look at the ones I'm wearing.
I look at my brother's too.
I can't wait till I am old enough
to learn what jeans can do.

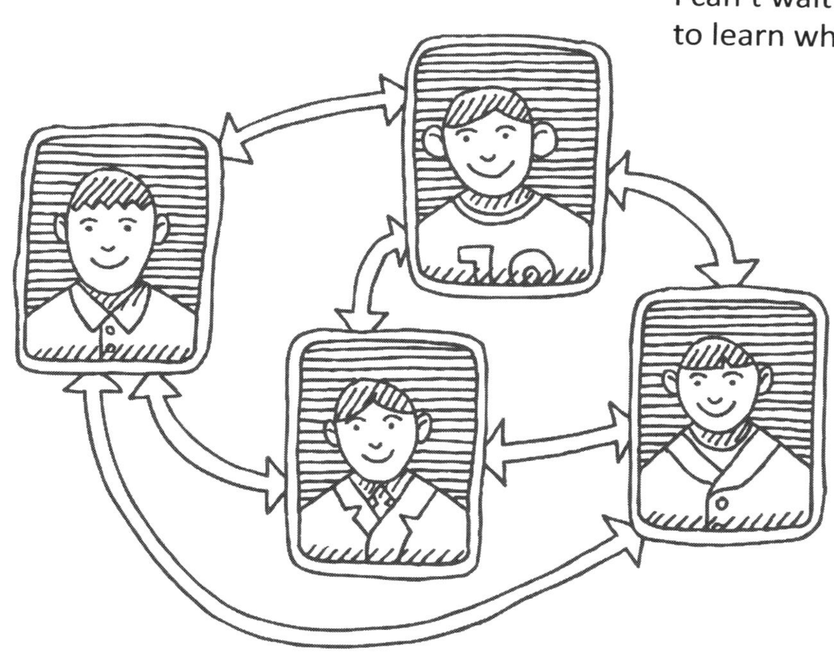

WONDERING WHY

Charles Darwin (1809-1882)
by **Shirley Smith Duke**

Darwin sailed as a naturalist on a far sea journey—
the *Beagle* was his ship.
He gathered samples of the different life he saw
from the Galapagos Islands off South America.

He took careful notes and then drew.

Different beaks on finches for each island.
Different shells on tortoises on the islands.
He thought and thought about why they looked alike,
yet had some noticeable differences.

Darwin had many questions.

Why do differing groups live in different places?
Do the strongest and best fit survive?
Can the young get their traits from their parents?
What makes species change over time?

Natural selection?

Darwin gathered up tortoises, fifty of them,
to take back to England to study.
But, as the ship sailed home, sailors ate turtle soup—
thank goodness for Darwin's drawings and notes!

Note: Darwin's theory of evolution by natural selection, controversial when introduced because the idea appeared to contradict the prevailing religious idea on how the world was created, has since become accepted by science.

LIFE STORY

by **Linda Ashman**

One hundred years of living,
and only this to tell:
slept a lot, grazed a little,
lugged this heavy shell.

Note: The Galápagos tortoise has the world's longest lifespan and is also the largest tortoise. These giants, native to the Galápagos Islands (600 miles west of Ecuador in South America), can live for more than 150 years. The males grow as long as 6 feet long from head to tail, and weigh more than 500 pounds. Most of their time is spent resting, grazing, and wallowing in water.

THE LAMENT OF LONESOME GEORGE

by **Jane Yolen**

"Lonesome George is gone and there will never be another like him."
—*New York Times* obituary, July 2, 2012

My name is George and I have lived
a long and simple life.
The last of all my tortoise kin,
no children and no wife.

My quiet archipelago
holds all my simple needs.
Just living to a hundred years
is reckoned in my deeds.

My keepers are my only friends,
as I close down my race.
"A Giant Tortoise" papers said
"extinction has a face."

But as I face extinction,
this one small truth I see:
You humans are a lot of you,
but I was one of me.

ROCKY RESCUE

by **Robyn Hood Black**

In the South Pacific,
Lord Howe Island has a tale
of how a giant stick bug,
thought extinct, might prevail.

"Land lobsters" as they're called
had lots of woe in store
when, back in 1918,
a ship wrecked on their shore.

Rats skittered from the boat
and found the black bugs tasty.
"They're gone!" the experts said. "Each one!"
—a conclusion that proved hasty.

For not so long ago,
some scientists, at night,
climbed a sea stack miles away
and found an awesome sight.

Look! The giant stick bugs!
They counted twenty-four.
Now with help from science,
there are many, many more.

TITAN IN MAN'S SEAWEED

(West Indian Manatees)

by **Michael J. Rosen**

Gargantuan sirenian,
you lone marine mammalian
who's totally vegetarian,
you sea cow some have called a mermaid,
your trusting nature has been betrayed
because you never were afraid
of motorboats or nets or oceans
fouled by waste that evolution
never meant for your slow motion.
Almost extinct—how can that be?
Is it our need to eat or vanity
or inhumanity, oh manatee?

Note: The title is an anagram of the animals' name.

A NEW DINOSAUR

by **Marilyn Nelson**

A newly discovered German shepherd-sized dinosaur
provides a wealth of new information on the evolution of
bone-headedness.
The dinosaur, identified from fossilized bone fragments
unearthed in Alberta, Canada,
is one of the earliest pachycephalosaur specimens
ever unearthed. Pachycephalosaurus
(from Greek *pachys-/παχυς-* "thick",
kephale/κεφαλη "head" and *sauros/σαυρος* "lizard"),
lived during the Late Cretaceous Period
in what is now North America.
Paleontologists have named the new creature
Acrotholus audeti: its genus *Acrotholus*
("thick dome"), its unique species called *Audeti*
because it was found on a ranch owned by Roy Audet.
Acrotholus audeti—"Audi" for short—
roamed about 85 million years ago.
She walked or ran on long hind legs,
lived in herds, defended herself with head-butting.

Or so says Science, continuing Adam's task
of naming the animals. Naming, understanding,
paying due awe to each trace of the evidence
of Papa Creation's unfolding, evolving design,
which is denied by the pachycephalic now among us.

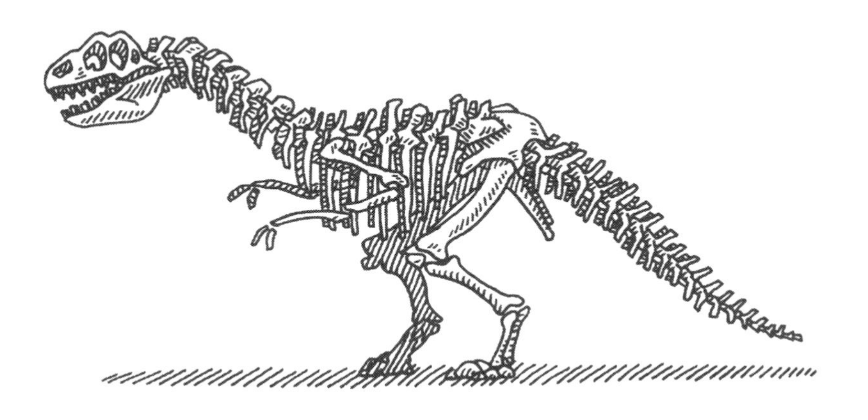

WHAT IS A FOOT?

by **Jane Yolen**

You will find a foot at the end of your limb,
Where you might wear a fin when you go for a swim.
It's got segments galore, it's got bones by the dozens,
And the bones have more bones, who are all sort of cousins.

As for animal feet, there's a soft foot, or paw,
That ends in strong nails, and is often called claw.
But others have hard feet, a hoof as we say.
And that is a feat of foot facts for today.

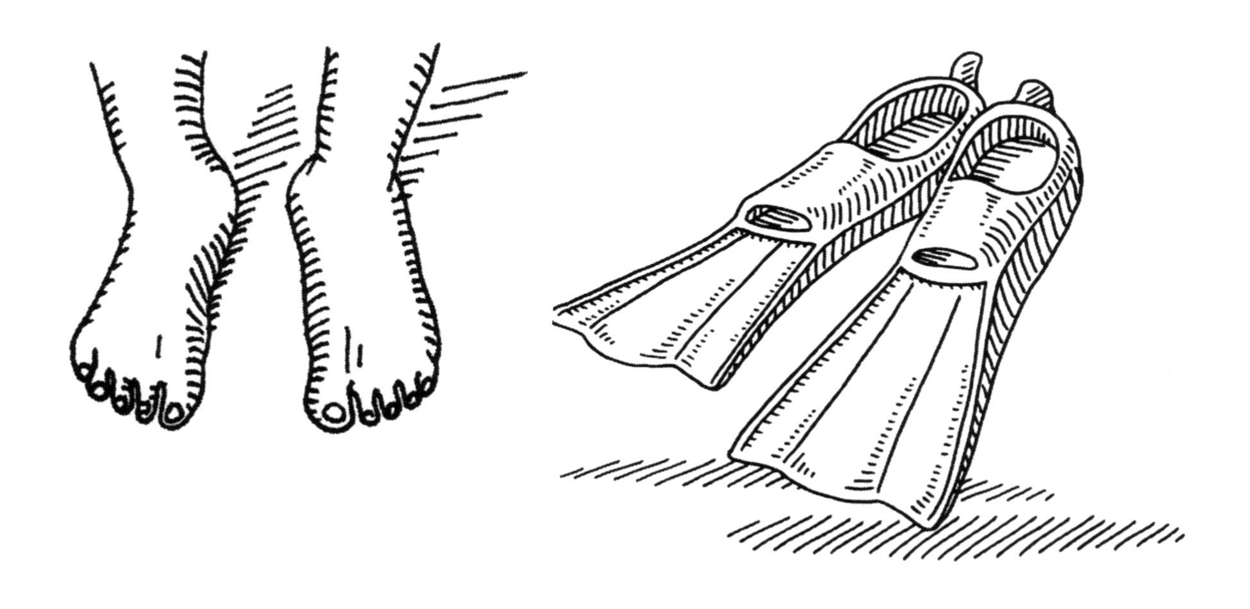

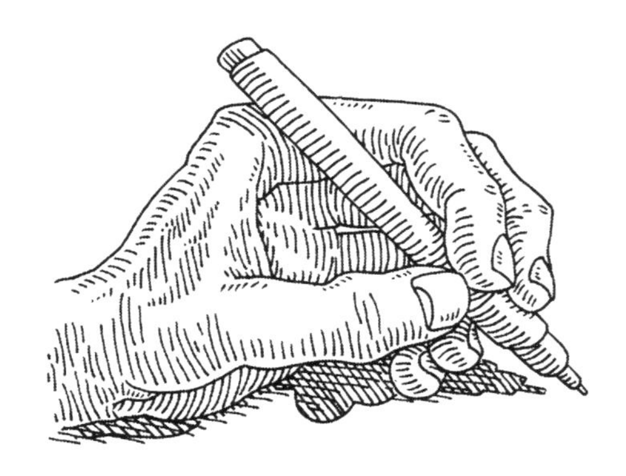

HANDS

by **Kate Coombs**

I bend my fingers,
move thumbs like this.
Knuckles, nails,
palms and wrists.

With my two hands
I can push-pull hard.
I can plant a flower
in our backyard.

I can catch a bug,
I can open a door.
I can pour the milk
and touch the floor.

I can hold the baby,
play a video game.
I can throw a ball
and write my name.

My devices are nice—
they can talk, beep, sing.
But my two hands
can do anything.

LET ME JOIN YOU

by **Heidi Bee Roemer**

Your
elbow
isn't cute.
It is wrinkly,
red, and rough. But just
try to live without it!
Simple tasks would sure be tough.

Your
elbow
joint is a
vertex. Mark it
with a B. Then draw
an A upon your hand.
Now let's tag your shoulder C.

With-
out your
elbow joint,
I wonder, how
would you touch your toes?
How would you brush your hair?
And how would you blow your nose?

Your
elbow
isn't cute,
but that ***vertex***
labeled B, keeps your
useful parts connected,
like your ***endpoints*** A and C.

C
B
A

ARMOR

by **Margarita Engle**

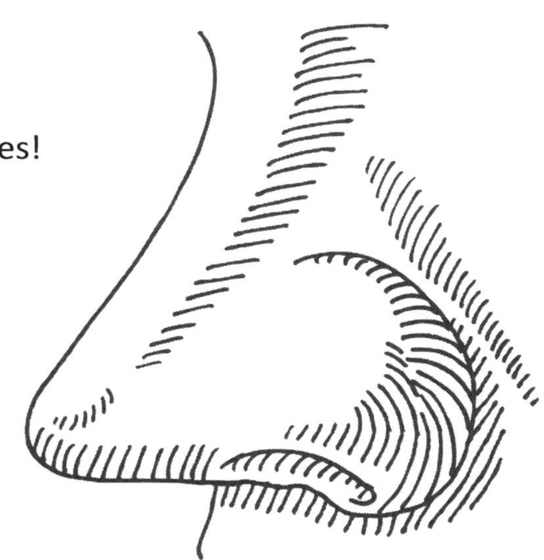

Jagged spikes, prongs, spears, barbs, spines!
Until I peered at tiny pollen grains
under a powerful microscope,
I had no idea that flowers
could be so fierce.

No wonder my nose feels
scratched and scraped
in hay fever season!

Note: If you look at a variety of pollen grains, magnified by an electron microscope, you might see protective spikes that keep pollen from being eaten by insects.

PROTECTING MY FRIEND

by **Jacqueline Jules**

Pollen in the spring
makes me sneeze.
But peanuts anytime
make Jillian wheeze.
Even a little bit
can make her sick,
so after I eat
I wash up quick.
And I don't worry
about being rude,
because it's safer
not to share my food.
That's how I
protect and defend
kids with allergies,
like Jillian, my friend.

SEEING SCHOOL

by **Kate Coombs**

My desk is in back
and I can't really see.
I ask to move up—
now it's glasses for me.

Don't want to wear them.
Do I have to, Mom?
But the world takes shape
when I put them on.

I see letters and words
all over the place,
numbers and edges
and my teacher's face.

I can see all the smiles
when I come to class.
Today I'm brand-new
with two pieces of glass.

DR. LEE

by **Janet Wong**

Last week
I couldn't
see the board
in any
of my classes.

Dr. Lee
saw
right away
that I just need
some glasses!

When I grow up
I want to be
a doctor
just like
Dr. Lee!

SHOTS! SHOTS! SHOTS!

by **Joy Acey**

I need the shots,
please make it quick.
I see a needle,
think I'll be sick.
You say it is just
a little stick.
I know vaccinations
build immunity
but getting them
takes bravery.

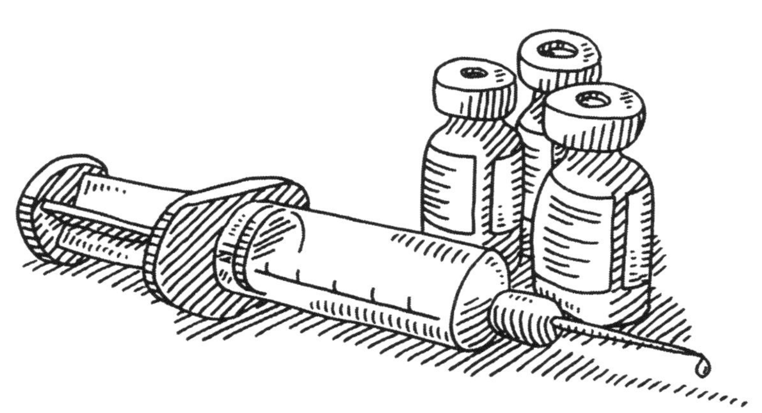

THAT DiSH THING

by **Virginia Euwer Wolff**

Richard Julius Petri,
a stout and sturdy German,
was interested in illnesses
and microscopic vermin.
That vile tuberculosis,
its victims aching, moaning;
he didn't know his research
would someday be called cloning.

Distributing bacteria
in 1882,
he cultured, peered, and poured and stirred
to find out something new.
But lab equipment way back then
was quite unlike today:
no perfect vessel for the cells.
He pouted in dismay.

The tubes, the flasks, their angles
wouldn't let bacteria thrive.
Without a way to grow them,
how could the search survive?
Oh, woe for Richard Julius
but, acting on his wish,
he devised a flat container
and now we use the Petri dish.

CANCER
by **Mary Lee Hahn**

Cancer's *what*
is cells growing wild.

Cancer's *who*
is man, woman, or child.

Cancer's *why*
is scientists aren't sure.

Cancer's *hope*
is someday a cure.

Cancer's *enemies*
are surgery and drugs.

Cancer's *helpers*
are flowers and hugs.

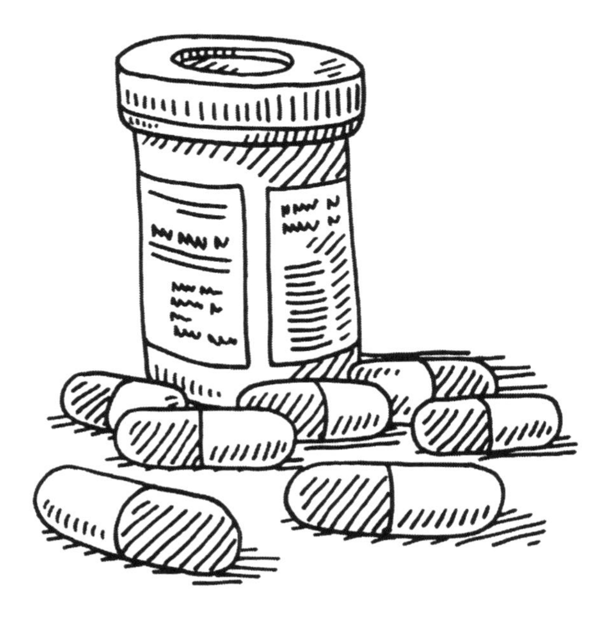

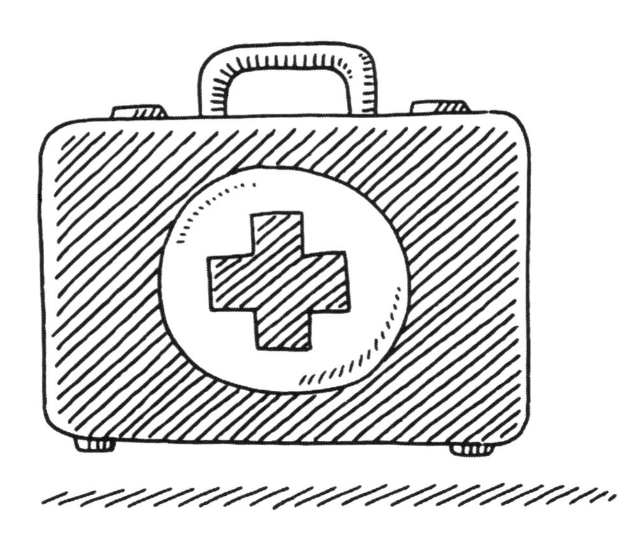

I WANT TO KNOW WHY
by **David L. Harrison**

I want to know why
People get sick.
I want to know why
Little kids die.
I want to help find
Cures for diseases.
Someone must do it.
Why not I?

I want to know why
People are starving,
Why kids suffer
Such misery.
I want to help find
Cures for famine.
Someone must do it.
Let it be me.

DA VINCI DID IT!

by **Renée M. LaTulippe**

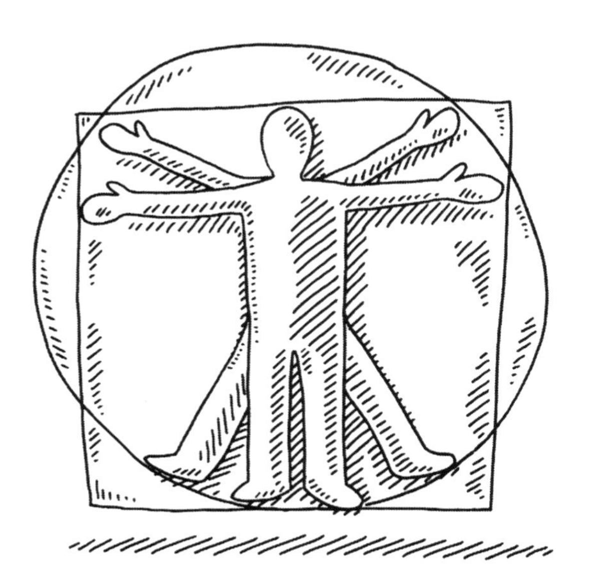

In Italy long, long ago,
a genius lived—
LEONARDO!

He was—
a painter, sculptor, mathematician,
engineer, and skilled musician

who dreamed up—
robots, carts, and parachutes,
flying planes and diving suits.

In fact—
as long as time did not forbid it,
you can bet da Vinci did it!

THE "BLACK LEONARDO"*

by **J. Patrick Lewis**

George Washington Carver
1846–1943
Botanist, educator, soil scientist, and inventor

He analyzed the peanut
And the sweet potato too,
Developing things like plastics, paints,
Linoleum, shampoo,
Peanut butter, vinegar,
Insecticide, and yeast.
"Sometimes," he said, "you find the
Secrets most among the least."

**Time Magazine* reference from an article published on November 24, 1941

THE ENGiNEER

by **Stephanie Calmenson**

Listen up and you will hear
Why I am called an engineer.

I solve. I build. I invent.
I'd say my time is very well spent.

Want a bridge? I'll design it for you.
Want a new kind of wheel? I'll develop that, too.

I use computers. I use my brain.
I think and test till the answer is plain.

Want a robot, a rocket, an electronic device?
I'll take the assignment. I won't think twice.

I'll make running shoes that will send you soaring!
I'll develop a device that will keep you from snoring!

My life is all about invention.
Making the world work better is my intention.

THE CRANE OPERATOR

by **Rebecca Kai Dotlich**

He knows which lever to pull
and how to lift that tree,
he's checked the size and weight;

those details are the key

to doing his job, and doing it right—
he raises and lowers
the boom day and night;

he figures the distance,
he studies the chart—
operating a crane

is construction art.

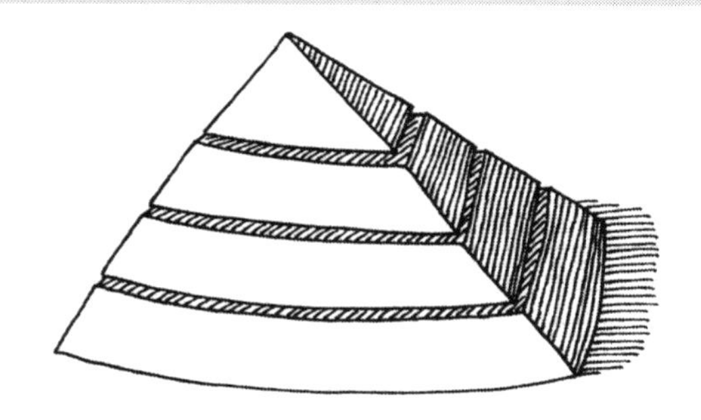

THE GREAT PYRAMID OF GIZA

(completed around 2570 BC)

by **Laura Purdie Salas**

Four thousand sandy years ago,
Egyptians built a sacred site.

They had
no steel,
no great machines,
no software help
to get it right.

They used a wooden "tool of knowing"
to read the sky, keep stars aligned.
They aimed a fragile palm leaf's spine
to keep all angles well-defined.

For 20 years, the workers toiled,
hour by hour and block by block.
They quarried limestone in the sun
and heaved out mighty granite rock.

No pulleys pulled, no big cranes raised.
With papyrus ropes, they dragged each stone—
two tons—then levered into place . . .
We think—so much is still unknown!

Made of rock, with wood and sweat,
built without one iron beam.
Giza held the record height
for centuries.

Now, that's extreme!

LEVERS

by **Michael Salinger**

A screwdriver
opening a big can
of house paint
is a machine
that is simple
and clever.

Just a beam
and a fulcrum
distributing force
and you've made yourself
a lever.

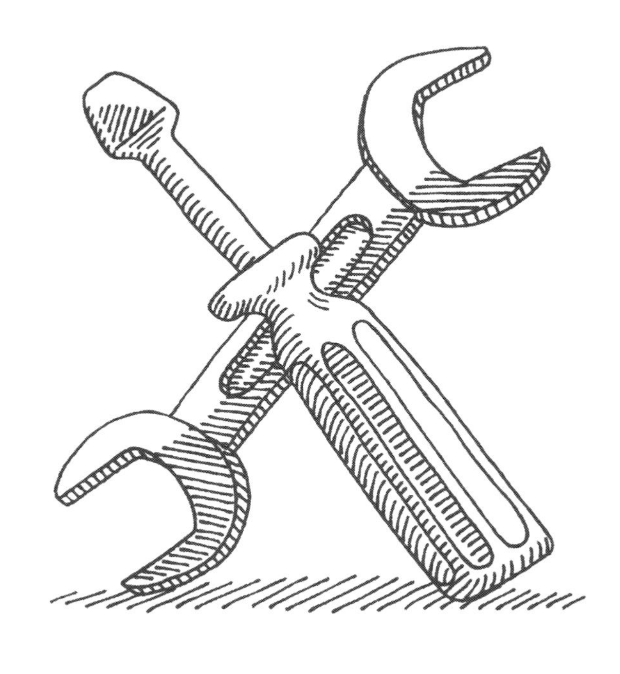

GEARS

by **Michael Salinger**

A gear is a machine
that needs only two parts.
Like wheels with teeth,
when one spins the other starts
to turn in what is called a ratio.
Gears come in all
different shapes and sizes,
mostly doing their work
inside of stuff.
Where might we use
some gears today?
What spins or turns?
What rotates or grinds?
What lifts or what lowers?
How many gears can you find?

FOUNDATION (DON'T RUSH IT!)

by **Charles Waters**

Where to begin?
These pieces are all chunked together.
Directions read
like a pile of hieroglyphics.
I cross my arms and stew.
Grandpa walks by.
"What's wrong?" he asks.
"I don't know where to begin," I reply.
"A good rule of thumb
is to start at the bottom,
have a good base.
That's called a foundation."
He reads the directions slowly as
we begin to slot pieces together
on the wooden floor.
"Being patient is key—remember,
once you rush it, you crush it."
This building starts rising like helium,
turning into the skyscraper of my dreams.

DRIFTWOOD HUT

by **Renée M. LaTulippe**

Today we'll build a driftwood hut,
my brother and I, we two.
A high-tide haul of branches means
that we have work to do!

We stack up thirty sea-slick sticks,
and figure that it's plenty.
But stacking sure is fun, and so
we stack an extra twenty.

The trick is lashing them. You need:
some twine, some time, a brother.
Two go up. Three fall down—
oops! Stack and lash another.

We wrap the hut with tattered tarp
and seaweed washed ashore,
then gather dune-grass, twigs, and shells
to pave our sandy floor.

Our driftwood hut is secret—thick
with salt-wind rushing through.
Just big enough to whisper in—
my brother and I, we two.

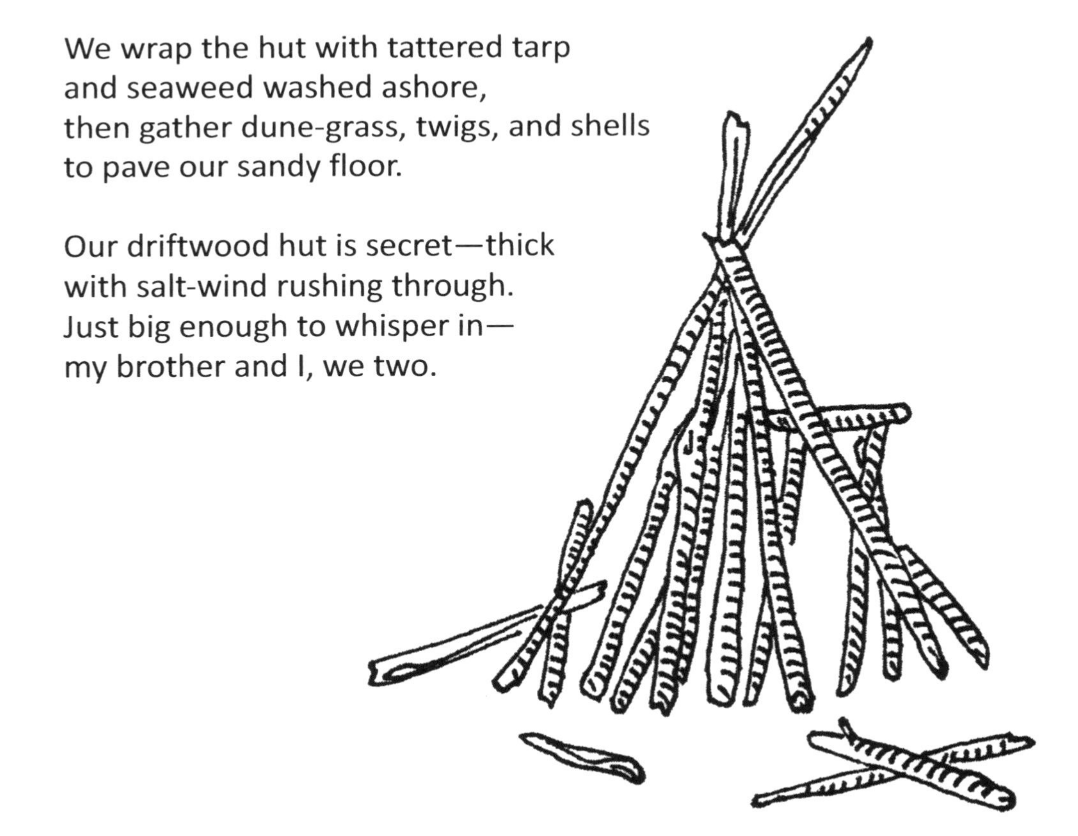

TINKER TIME

by **Janet Wong**

In Grandpa's basement you can find
gears and wheels and wire and twine,

lots of nuts and bolts and hooks
and one whole shelf of build-it books.

If we need help during Tinker Time,
we go to the computer and look online.

What will we build when we're all done?
We don't know yet—that's half the fun!

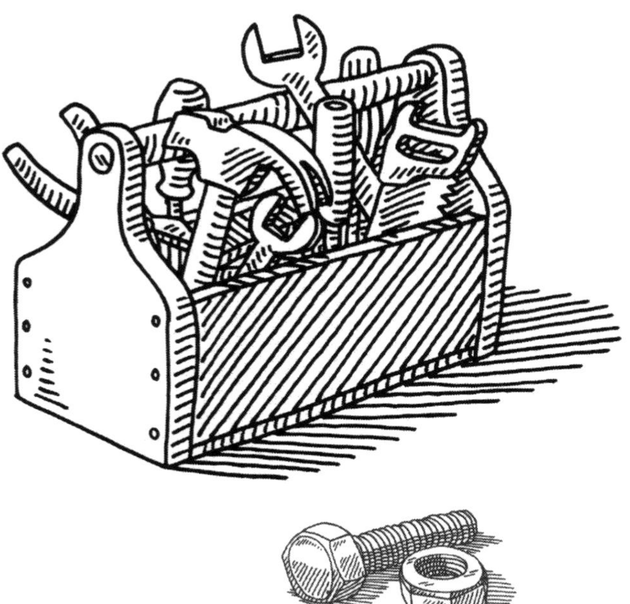

I THOUGHT I BUILT A DOG HOUSE

by **Eric Ode**

I thought I built a dog house;
a house where dogs belong.
I thought I built a dog house.
I must have built it wrong.

I'm certain if you saw it,
you'd see that it's too small.
My puppy's tail can hardly fit.
His head won't fit at all.

I thought I built a dog house.
I hung it in a tree.
I guess I built a bird house,
and that's just fine with me.

MY ROBOT

by **David L. Harrison**

I built my first robot with sticks and mud.
I gave it a pebble brain.
Turned out to be a dud.
You might say that experiment was in vain.
It dissolved when I left it in the rain.

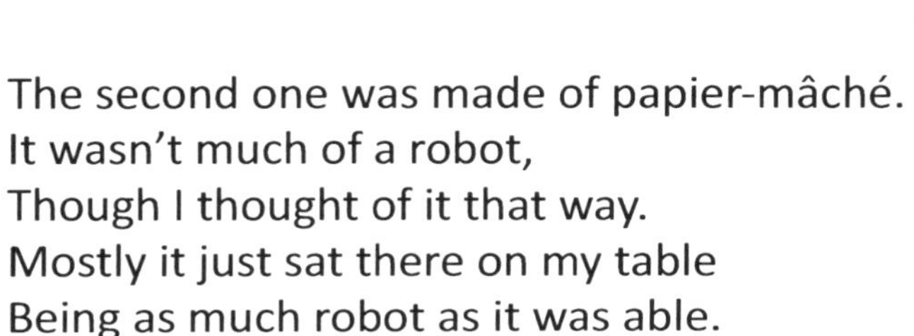

The second one was made of papier-mâché.
It wasn't much of a robot,
Though I thought of it that way.
Mostly it just sat there on my table
Being as much robot as it was able.

The third robot I built was from a kit.
It had a metal frame
And a motor came with it.
It looked really truly like a robot.
Even so, it couldn't do a lot.

One of these days I'm going to get it right.
The robot in my future
Will be a glorious sight.
My robot will follow me around.
My robot! What an awesome sound!

METAL MONSTER

by **X.J. Kennedy**

Mother lets out a loud cheer when
Our vacuum-cleaning robot
Purrs swiftly through our living room,
A metal monster. Oh, but

She sits and munches chocolates,
Watching it clean our rugs.
And when that sweet machine is done,
She gives it heartfelt hugs.

SODA MACHINE BITE

by **Jacqueline Jules**

Five rows of plastic bottles,
lined up in bright colors
like alphabet blocks.
All I have to do is
feed two bills into
a metal mouth with rollers
and press button
A5 for lemon-lime.
Wait!
It's B5 for lemon-lime,
and A5 for cola.
Too late!
The dispensing coil
punches the wrong bottle
like a bulldozer
down to the drawer by my knees.
That's not the drink I wanted,
but my money has been scanned
by optical sensors and swallowed.
This is the drink I get.
Might as well unscrew the top
and enjoy!

WHAT SHOULD I CALL IT?!

by **J. Patrick Lewis**

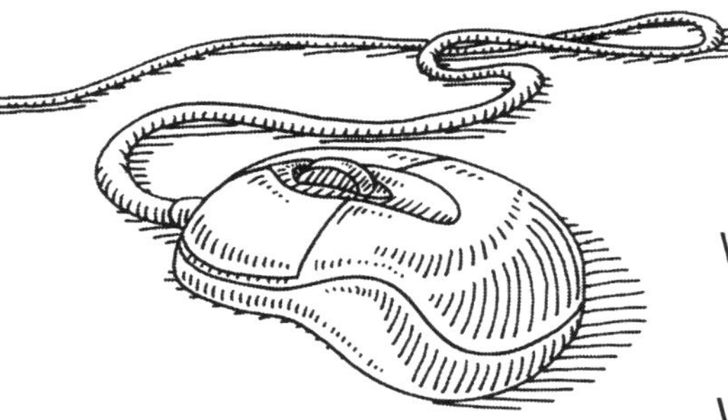

I was typing on my keyboard—
An old-fashioned A-to-Z board—
When a mouse hopped into my computer drawer!
And my nimble brain exploded,
Like the Big Bang years ago did,
With ideas no one's ever had before.

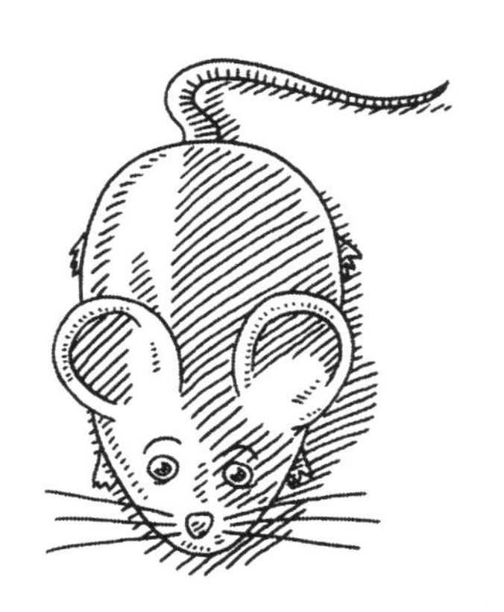

Could I change what I was writing
With an instrument by sliding
It across a pad as quickly as a wink?
Could I lift a word, erase it,
Move it here or there, replace it
With the *perfect* word? I had to stop and think!

Let me see, what should I call it?
Mole or chipmunk? Bat or ball? It
Moves an arrow back and forth across the screen
With one finger. *Mama Mia!*
That gives me a bright idea.
Help me name this handheld rodent-sized machine!

Note: Doug Englebart (1925-2013) invented the computer mouse in 1963. A mouse did not hop into his computer drawer; he called it a mouse because the wire that came out of it looked like a tail.

PRINTING, PRESSED BEYOND WORDS . . .

by **Robyn Hood Black**

Our printers today are still evolving.
So many projects—and problems they're solving!

In layers of plastic, a virtual mold:
printers are spitting out things you can hold.

These 3-D devices can also print gels,
stacking amazing assortments of cells.

Need a blood vessel? An organ, an ear?
Bioprinting is real—bioprinting is here!

COMPUTER MODELS

by **Janet Wong**

We have an engineer
visiting our classroom.
She shows us
how she uses her computer
to test designs:
Plan A,
Plan B,
Plan C.
She doesn't
have to build
a whole building
to see if it's better
to have six floors or seven or eight.
She can calculate
how much steel and glass
each different plan
would need,
how much it heat
it would use and lose,
how many hours
of construction time
and the cost.
She can tell us everything
about the plans, and—
click-click-click-click-click!—
with her camera
and five minutes
of cut and paste
and Photoshop,
she can put us
inside her buildings,
waving hello!

GAME PROGRAMMER

by **Janet Wong**

My aunt has the best job ever.
She programs video games.
Someone else
comes up with the stories.
Someone else
comes up with the names.
But she puts all the commands in.
She makes the games work right—
so cars will move
when you want them to,
so soldiers can see at night.
She speaks to the computer
and calculates the scoring.
My aunt has the most incredible job—
it's never, ever, boring!

COMPUTER GEEK

by **Carmen T. Bernier-Grand**

Draw an animated creature
Make it move in a game
You'll be a Game Designer
I'll play your games.

COMPU-NERDO

por **Carmen T. Bernier-**

Dibuja una caricatura animada
Hazla mover en un juego
Serás diseñador de juegos
Jugaré tus juegos.

MMO

by **Anastasia Suen**

At any hour
of any day
thousands of people
go online to play.

In the game
you are not yourself.
You can play as a wizard,
a knight, or an elf.

Can you work on a team
and communicate?
Help your clan save the world
before it's too late!

At any hour
of any day
thousands of people
go online to play.

Note: MMO is short for MMORPG, a massive multiplayer online role-playing game.

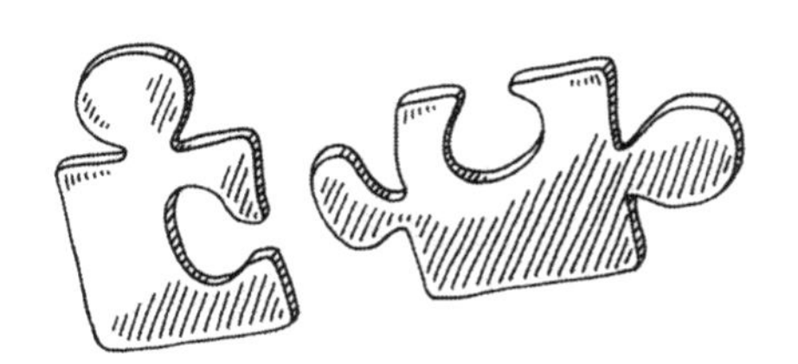

PIECES
by **Renée M. LaTulippe**

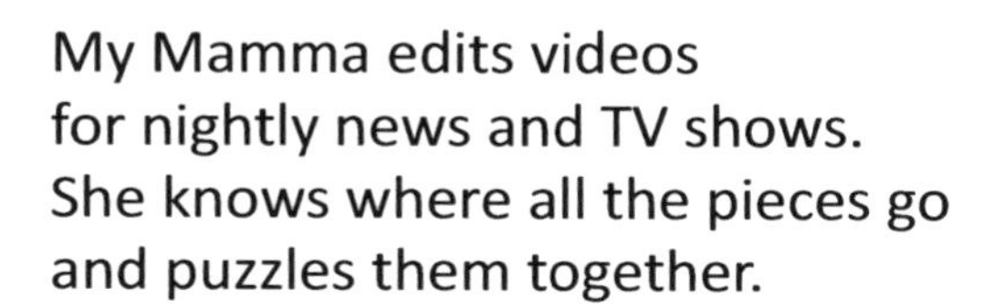

My Mamma edits videos
for nightly news and TV shows.
She knows where all the pieces go
and puzzles them together.

She often lets me see the screen
and watch the frames go, scene by scene.
A piece of person, sky, and tree:
we puzzle it together.

VIRTUAL ADVENTURE
by **Renée M. LaTulippe**

Yesterday I scaled some peaks.
Looky here: wind-chapped cheeks!

Right after lunch, I rode a gnu,
caught cuckoo birds in Katmandu.

Snowboard? Check. Windsurf, scuba.
After dinner? Played a tuba.

Safari in the Serengeti,
tango with a sweaty yeti.

I can do most anything—

from biking in downtown Beijing
to wrestling deep-sea squid-eos—

with my green screen videos.

FRAMES PER SECOND (FPS)
by **Janet Wong**

I'm trying to do animation
for my own video game.
I need to divide each action
into pieces, frame by frame.
I wonder: how many frames?
I don't want it to look rough.
I do a search on the Web
and read some shocking stuff:
most video games use
thirty frames per second (fps)—
some use much, much more,
some a little less.
So now I have an idea,
a smarter, better one:
I'll outsource all the work—
and I'll just play for fun!

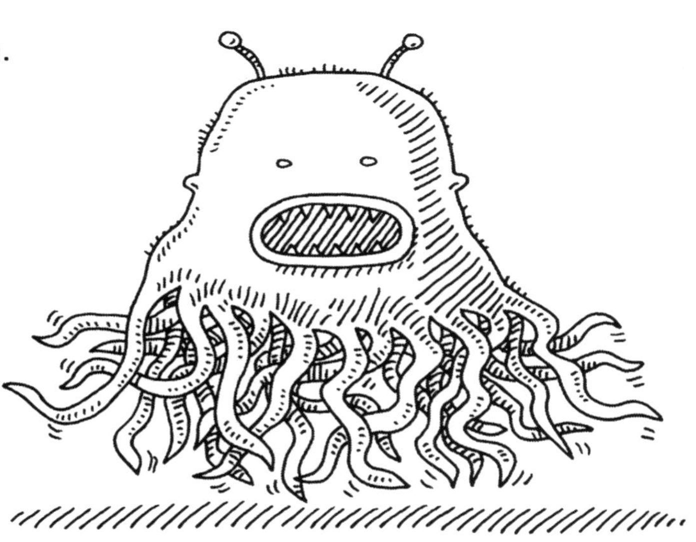

NURSING MATH

by **Jeannine Atkins**

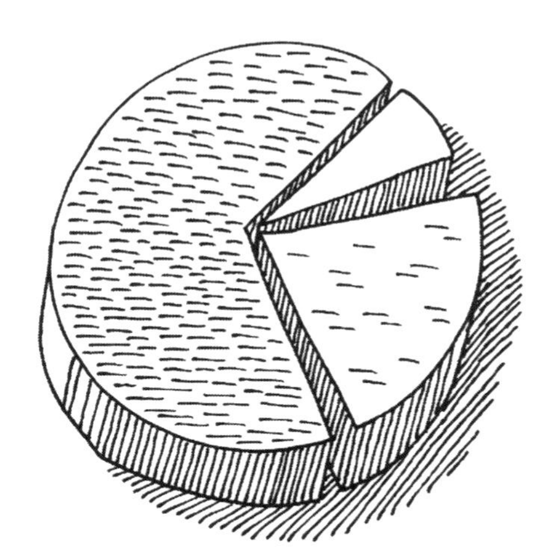

Florence Nightingale measured medicine and bandaged
arms and legs. She drew charts showing who got strong
and which soldiers stayed sick too long. She saw that
hospitals must be made cleaner and make other reforms.

How could she make this clear to Queen Victoria,
whose eyes glazed at numbers and words? She'd calculate!
The good nurse studied statistics, then drew columns
and charts, with lines curvy and straight.

At the castle, she curtseyed, then showed a circle
divided by lines and with labels. The queen found
it easy as pie to see why change must come, and it would,
from a mathematical nurse and a woman wearing a crown.

Note: Florence Nightingale (1820-1910) is famous as the founder of modern nursing. She also developed ways to use diagrams and graphs, creating a forerunner of what is now known as a pie chart.

MY BEAN PLANT

by **Amy Ludwig VanDerwater**

I made a graph so I can show
how every week my plant will grow.

It's planted in a paper cup
and every week my graph goes up.

I can tell you what this means.
I am good at growing beans.

STOPWATCH
by **Janet Wong**

We start timing
with a click—

then

tick tick tick tick tick tick tick tick tick tick
tick tick tick tick tick tick tick tick tick tick
tick tick tick tick tick tick tick tick tick tick

we click again.

Thirty seconds just passed?
Oh my!
Our stopwatch
counts *so* fast!

MOVING FOR FIVE MINUTES STRAIGHT
by **Betsy Franco**

For 50 secs,
we hop around,
then switch
for 50 more.

For half a minute
we all lie down
for push ups
on the floor.

For 90 secs,
we get in rows
to do our
jumping jacks.

For 80 secs,
we're on our feet,
all sprinting
up and back.

But when we
hear the buzzer sound,
we freeze.
We're glad to stop.

My pulse:
one hundred sixty-two!
I think my heart will
pop!

CRAZY DATA DAY

by **Janet Wong**

We rolled a car down the ramp
four times
and timed it with my stopwatch.

Six seconds.
Six seconds.
Six.
And six.

And then
the car
started playing tricks!

The next time was
NINE!

I cleaned the wheels.
George checked the track.
We rolled again.
Did the six come back?

What?
THREE!
No way!

What a Crazy Data Day!

ZAPPED!

by **April Halprin Wayland**

Dad says
that a penny and a galvanized nail
can produce an electrical current in fruit.

In *fruit?*
Hmm . . . am I doing this right?
Wait . . . *WOW!*

Now my heart's racing—
I'm filling pages:
how much voltage did each fruit produce?

Tomato: .59-.62
Kiwi: .85-.86
Orange: .89-.90
Lemon: .91-.93
Grapefruit: .93-.94

(Sometimes science makes me hungry.)

I'm so jazzed,
if you measured *my* current
on the voltmeter,

you'd be shocked.

GOING BANANAS

by **Heidi Bee Roemer**

Monkeys are competing on the beach beneath cabanas.
Going for the gold, five monkeys gobble up bananas.

Judges track the data, jotting down each monkey's score,
till the monkeys at the beach can eat not one banana more.

Judges post the totals of bananas each consumed:
10, 10, 18, 20, and a whopping 42!

Add up the bananas in this dandy data set.
Divide by 5 to find the ***mean*** and 20's what you get.

The ***mode*** refers to figures that appear once—and again.
The most repeated number in this data set is 10.

In math, position matters. Like a "monkey in the middle,"
the ***median*** is 18 in this silly data riddle.

THIRSTY MEASURES

by **Heidi Bee Roemer**

I pour some juice into my **cup**.
8 fluid ounces I drink up.
That's 16 tablespoons. Good stuff!
But the juice is not enough.

I mix a **pint** of lemonade,
the pink kind that my grandma made.
That's 16 ounces, 2 cold cups—
The lemonade is not enough.

I chug a **quart** of chocolate milk
and not one drop of milk is spilt.
That's 32 ounces, 4 cold cups,
2 pints of milk. Still . . . not enough.

I swig a **gallon** of iced tea.
That's 16 cups of tea for me.
I reach my peak capacity . . .
8 pints, 4 quarts, all gone. Oh no—
Excuse me, please. I gotta go!

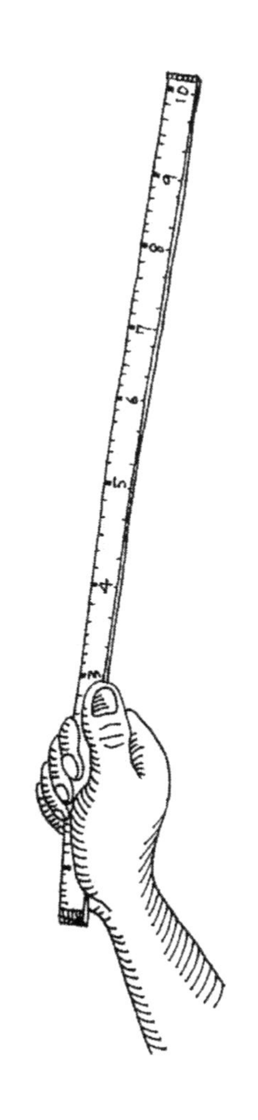

METER STICK

by **Amy Ludwig VanDerwater**

It's a pleasure to measure in meters.
It's a pleasure to measure because
everything measures in units of ten.
It measures so sweetly.
It does.
Ten centimeters are one decimeter.
Ten decimeters, a meter.
Divide.
Multiply.
Always by ten.
Measuring couldn't be neater.
And when I must measure
a plant or a pencil
when I must measure
a scrap of my day
I am connected
to all those who measure
in meters
in countries
so far
far
away.

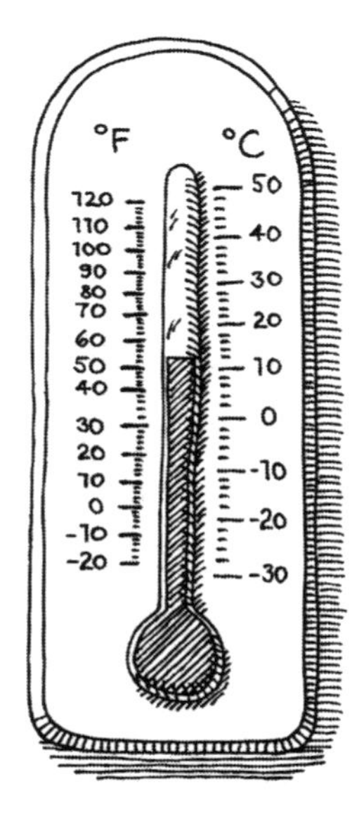

CELSIUS THERMOMETER

by **Renée M. LaTulippe**

The teacher taught us Celsius
to measure temperature.
She showed us the thermometer
and where the numbers were.

I learned that water's freezing point
is right at 0 degrees.
100 makes the water boil—
Celsius is easy!

PATTERNS IN NATURE : NATURE IN PATTERNS

by **Shirley Smith Duke**

patterns in nature	nature in patterns
bilateral symmetry	symmetry bilateral
middle a line	line a middle
halves in two	two in halves
matching exactly	exactly matching
one makes another	another makes one
identical images	images identical
—exactly alike—	—alike exactly—
mirror opposites	opposites mirror

3-D

poem for two voices
by **Betsy Franco**

A rectangular prism can be a box
from a whirlwind shopping spree.

A pyramid in Egypt is
giant geometry.

A cone can be
an ice cream treat on a sizzling summer day.

A cube can be
the game board dice when you're playing inside 'cause it's gray.

And . . .

A truncated icosahedron
is a regular soccer ball.

Take a careful look at the shape you kick
while practicing next fall!

Note: The truncated icosahedron is created by truncating (cutting off) the tips of the icosahedron one third of the way into each edge.

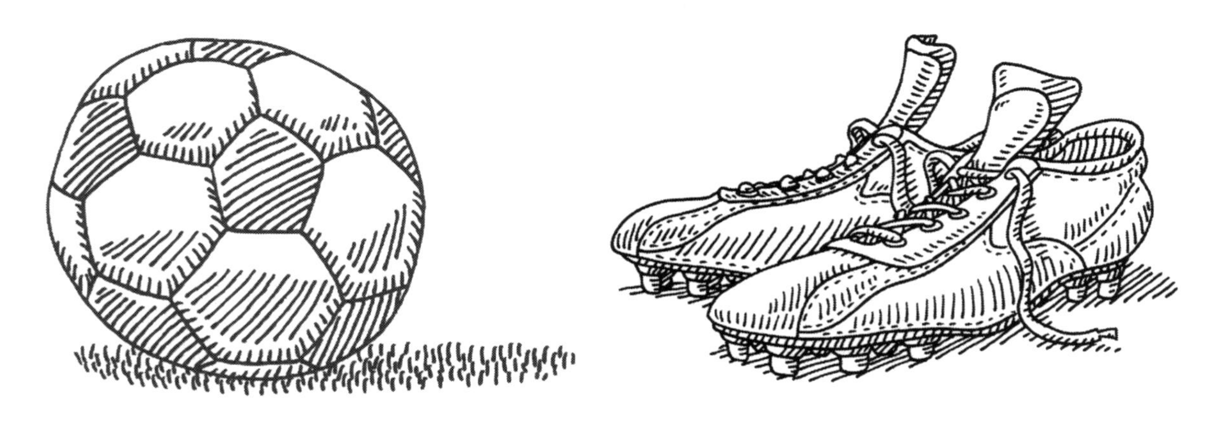

THE REAL THING

by **Linda Sue Park**

They say we're getting closer
to a real invisibility cloak.
It's not ready yet,
but it will be, someday soon.

Would you want one?
Would you try it out at the store,
beg for one for your birthday,
put it on your wish list?

The first ones for sale
will be really expensive.
But they'll come down in price
a bit at a time

until maybe you could save
(allowance, babysitting, lawn mowing)
for the basic model—the IC 100,
which doesn't quite cover your shoes.

It wouldn't take long
before lots of people had them.
There would be rules
about not bringing them to school.

New sports leagues!
IC basketball. IC soccer.
It sounds like 'I See,'
but of course, you wouldn't,

except for odd ripples:
the ghost of a thumb,
a flicker of cleats,
an eyebrow floating in midair.

FUTURE DREAMS IDEA #63

by **Janet Wong**

What if
you could make
a pillow-backpack

and when you touched a button,

it would pop up into
a Giant Air Shell
to keep you safe in an earthquake?

Mom asks what I'm doing
with my old broken backpack,
a roll of foil,
duct tape,
all our plastic bowls,
a whole bag of marshmallows,
my *sticky fingers*,
and her tablet.

I say, *Research!*

MY WRISTROBOT PACK

by **Carmen Tafolla**

I put my WristRobot on my wrist every day
As soon as it wakes me with my breakfast tray.
It lays out my clothes, all clean and pressed,
Checks out the forecast, and helps me get dressed.
Then it reminds me that my hair's sticking up,
Makes my bed, packs my lunch, and refills my cup.

It sets all the vectors to beam me to school
And makes sure I travel through Rome, Istanbul,
Paris and Bogotá on my way,
to guarantee I have an interesting day.
With Laser-Lev games and Photoelectronic Tag,
I can play with my friends wherever they're at.

I'd NEVER be without one. I'll always have it near.
—Wait a minute! Where am I? What's happening here?
Whose messy bed is this? Why no Laser-Lev games?
Mom, is this MY home? Was this all a dream?
NO WristRobot Pack?? They don't even EXIST?!!
...
Well then, I'll just make one—it's first on my list!

Sign me up for Physics and Electro-robotics.
I need Laser Science and Transmitter-crionics.
I want to study Electromagnetic Levitation,
Bilocation Engineering, Locomotion Actuation.
I'm really missing my old WristRobot Pack—
I'll invent it and THEN, get my DREAM life back!

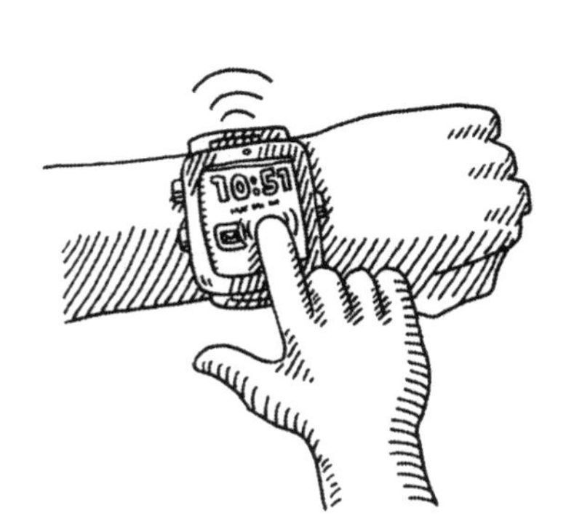

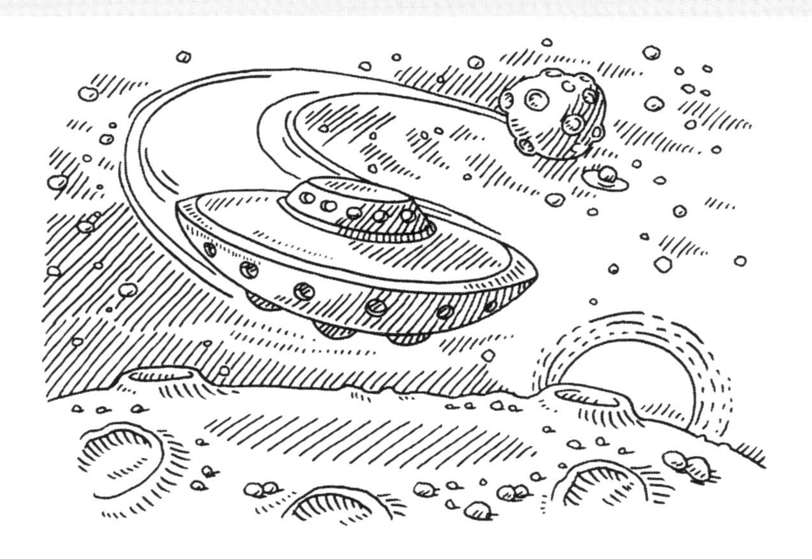

I PLAN TO BE AN ASTRONAUT

by **Kenn Nesbitt**

I plan to be an astronaut
and fly through outer space,
in search of distant planets
in some undiscovered place.

I'll travel through the galaxy
exploring near and far,
I'll analyze each asteroid
and study every star.

I'll navigate through nebulae
and circle every sun.
I'll cruise through constellations
on an interstellar run.

I'll scrutinize each satellite
from Saturn to the moon,
but never be gone longer
than a single afternoon.

For though it sounds like lots of fun,
I'm sure, for what it's worth,
come time to use the restroom
I would rather be on Earth.

EVERYDAY ASTRONAUT

by **Carmen Tafolla**

When they said that they were looking
for everyday astronauts
people from just normal life
Teachers with totes filled with things to surprise
Bus drivers with steady and focused dark eyes
Students peeking out from under stacks and reams . . .

I started to dream.
To see myself floating in space
waving at friends
from a moon away
a planet my nightlight
shining soft, night and day

Our little blue planet
is a warm, cozy room
in this wide and wonderful
wild and wakened
universe
we call home

UN ASTRONAUTA COMÚN

por **Carmen Tafolla**

Cuando oí que buscaban
astronautas ordinarios
entre la genta
de vida sencilla . . .

Imaginé maestros con mochilas
llenas de sorpresas,
Choferes de camión
con ojos de aguilón,
Jóvenes muy normales
con sus juegos digitales,
Mamás amorosas
Con caras de rosas.

Mi mente se subió
a una astronave suave
Mis fantasías y sueños
comenzaron a volar
Ahora floto en el espacio
saludando amigos
desde una luna lejana.
Mi lucecita nocturna es un planeta azul
que constante brilla, de noche y de día,
Mi habitación pequeña es el sistema solar,
un rinconcito cómodo de esta casa grande,
enorme universo,
que llamamos hogar.

SPACE YACHT

by **Juanita Havill**

Imagine a space ship with sails,
a solar yacht that travels far
and carries you from earth to Mars
on streams of photons from our star.

Never-ending streams of photons,
particles in waves of light,
bounce off sails to push your space ship
like blowing wind upon a kite.

Wind-powered kites attached to string
guided by humans on the ground
can never fly above the clouds.
Not so your yacht—it's universe bound.

Note from Space.com: "The largest solar sail ever constructed is headed for the launch pad in 2014 on a mission to demonstrate the value of 'propellantless propulsion'—the act of using photons from the sun to push a craft through space."

MOVING TO ATLANTIS CITY, 2112

by **Steven Withrow**

Eleven billion side by side
Take up a lot of space.
We needed fresh solutions quick
To house the human race.
We first built floating towns as big
As islands on the sea.
But soon these grew too overstuffed
For Mom and Dad and me.
We'd watched a holo on the web:
New Lab Needs Volunteers!
Our family signed up to be
Among the pioneers.
Of all the kids they picked to live
Beneath the ocean's rim
I wasn't quite the perfect choice.
I couldn't even swim.
But once we rode the shuttle down
And passed the pressure sphere
I knew there was no surface place
So much like Home as here.
Although I sometimes miss the sun
And unrecycled air,
I'd never trade my deep dark view
With all of you up there.

INVENTION INTENTIONS

by **Kristy Dempsey**

Welcome to my interview.
I have some questions (just a few),
designed to help me understand
which products might be in demand.
I'm striving for development
of something you'll find relevant:
what do you need, what do you wish
would make your life more simple-ish?
What products do you use the most?
Would you buy pre-buttered toast?
What's more important: cost or ease?
What would you pay me, if you please?
Is there an item you can't stand?
I'll redesign it with my plan!
My product just might fill a niche
Or maybe it would scratch that itch
The one you just can't reach in back . . .
My invention might be what you lack!

WHAT IS SCIENCE?

by **Cynthia Cotten**

Science is
knowledge
and
a process

Science is
how and why (or why not)
experimentation and observation
expected results and accidental discoveries

Science is
questioning
exploring
understanding

Science is
useful
exciting
ongoing
challenging

Science is
the study of
elements, compounds, cells
motion, sound, light
and of
the ocean,
the earth,
the sky,
and all that they contain

With all that science *is*
perhaps the question should be
"What *isn't* science?"

FUN WEBSITES FOR SCIENCE LEARNING

Here are some websites that provide helpful information and fun things to help you learn more about science!

AllAboutBirds.org
Animals.NationalGeographic.com
AnimalSpot.net
ArborDay.org
AustralianMuseum.net.au
BrainPop.com
Burpee.com
Calm.com
Census.gov
Chem4kids.com
CicadaMania.com
CloudAppreciationSociety.org
CompostingCouncil.org
ConserveTurtles.org
DiscoverE.org
DrawingsofLeonardo.org
EarthPopUpBook.weebly.com
EyeChartMaker.com
FamousScientists.org
FreeSound.org
FrisbeeDisc.com
GalaxyMap.org
GetCaughtEngineering.com
Glogster.com
HealthyPet.com
HeavyEquipment.com
HistoryofBridges.com
Howcast.com
HowStuffWorks.com
Illusions.org
IntoTheWind.com
Inventions.org
JaneGoodall.org
Kids.NationalGeographic.com
KidsButterfly.org
KidsHealth.org
Mission-blue.org
NationalGeographic.com
NASA.gov
NobelPrize.org
Optics4kids.org
OrganicGardening.com
Panda.org
Papertoys.com
Photography.NationalGeographic.com
PhotosynthesisForKids.com
RachelCarson.org
RocksForKids.com
ScienceBuddies.org
SmithsonianEducation.org/Scientist
SolarEnergy.org
SoundCloud.com
Space.com
SpaceCamp.com
TeacherVision.fen.com
TheReptileReport.com
Tree-Pictures.com
TryScience.org
Tsunami.org
Video.NationalGeographic.com
Water.org
Weather.com
WebElements.com
Windustry.org
WorldFishingNetwork.com

A MINI-GLOSSARY OF SCIENCE WORDS

Here is a short list of key science words used throughout the poems and their definitions.

Acceleration: To speed up and go faster and faster

Aerodynamics: The design of objects to allow them to move easily through air with reduced friction

Animation: The motion of characters on video created by a series of slightly changing individual images

Atom: The smallest bit of matter that still holds its own chemical properties

Avian: A term referring to birds

Bacteria: Single-celled microscopic life forms having no nuclear membrane holding genetic material

Beaker: A glass container with a straight side for measuring volume in laboratories

Biodiversity: The wide variety of all life forms found on the earth

Bioprinting: Creating new, living tissue from cells in a 3-D process in labs

Budding: The small bumps on a tree that grow into new leaves, shoots, or flowers

Calories: A measure of the amount of energy held in food

Camouflage: Natural protective coloration on living organisms that helps them blend in with their surroundings

Capillary action: The upward force of a liquid due to water molecules' tendency to stick together in tight spaces

Carbs (carbohydrates): The nutrients in food that supply energy; also a short name for *carbohydrates*

Casts: The kind of fossil formed when a mold is filled

Celsius: The scale for measuring temperature based on 100, with 0 for freezing and 100 for boiling

Chlorophyll: The green substance in plants that captures sunlight to harness its energy to make food

Cicada: A group of insects having a life cycle that varies in the number of years it appears, noticeable for the loud buzzing by the males

Cinder cone: The most common kind of volcano, and one that forms around a central vent or column in which the lava flows down the sides

Compost: A nutrient-rich material formed from decaying organic matter like leaves, vegetable parts, and manure

Cretaceous: A geologic period of time dating from 140 million to 65 million years ago during the time dinosaurs developed and then became extinct

Deforestation: The removal of all trees in a particular region or area

Drag: The force caused by the resistance of the air to an object moving forward

Drought: A period of time with little or no rain

Eco-agricultural: An integrated ecosystem approach to agriculture that will sustain rural farming, conserve biodiversity, and develop sustainable farms

Electromagnetic spectrum: The range or scale of radiation from the sun in decreasing wavelength size, such as radio, microwaves, visible light, and x-rays

Element: Matter that cannot be broken down into any other substance and still keeps its same properties

Elliptic: The shape of a flattened oval

Engineer: A person who is specially trained to design and build machines or buildings and create solutions to problems

Evaporation: The process in which molecules in a liquid move into the air as a result of heat speeding up their motion

Exposure: Film coming into contact with light to produce a picture

Flask: A specially shaped container with a narrow neck used in science labs

Fulcrum: The point on which a lever rests and turns about

Fungi: The plant-like organisms that take nourishment by growing on other plants or decaying material; singular is *fungus*

Gall bladder: A pouch-like body organ under the liver that secretes bile to help digest fats

Gear: A toothed wheel that uses a change of direction to turn other toothed wheels to determine the rate of motion

Genes: The hereditary material in the nucleus of cells that determines characteristics for developing cells

Genetic: Having inherited traits or characteristics from one's parents

Geologist: A scientist who studies the earth and its processes

Graft: To insert a cutting from one plant into another plant so that they grow together as one plant

Graph: A chart or diagram that shows the relationship of one thing to another in numbers or amounts

Gravity: The attractive force between two objects that pulls the smaller one toward the larger one

Greenhouse gases: Carbon dioxide and other gases that trap extra heat from the sun and warm the earth's atmosphere

H_2O: The chemical symbol for water signifying two atoms of hydrogen bonded with one atom of oxygen to form water

Humidity: A high amount of water vapor in the air at a given time

Humus: Rich soil formed by the decay of plants and animals

Hybrid: An automobile that runs on both gasoline and electrical power from a battery

Hypothesis: The knowledgeable prediction an experimenter states in order to design an experiment to learn if the answer to a question is accurate

Immunity: The body's ability to build up defensive cells that prevent illnesses caused by germs

Indigo: A dark purplish-blue color on the visible spectrum

Lever: A simple machine made of a bar and a fulcrum around which the bar pivots to lift loads

Lift: The force that moves an airplane upward caused by the forward motion of the plane and its wings

Molecule: The smallest bit of matter from a compound (made of two or more elements) that still retains the properties of that matter

Natural selection: The concept that all species evolved from common ancestors and that the best and most fit survived to pass along their characteristics while the weaker ones died out

Naturalist: A person who studies nature and its science

Near Earth Object (NEO): The comets and asteroids affected by the gravity of other planets that enter into orbit around the earth

Nebula: The cloud of dust and gas visible in space as a bright area at night

Niche: A place in the environment and an organism's role there

Nymph: The young form of some insects in which the young resemble the parents and shed their outer shell multiple times to fit their growing body

Observation: The information gained directly using the senses or exact measurements

Organism: A living thing like a plant or animal

Outsource: To get products from a foreign or outside supplier

Pachycephalic: A kind of skull where the bone is extra thick

Paleoentologist: A scientist who studies life that lived during ancient or geological times

Pancreas: A body organ that regulates the amount of sugar in the blood

Photons: Tiny particles that make up light waves

Photosynthesis: The food-making process carried out by green plants using the sun's energy, water, and carbon dioxide

Plankton: Microscopic, single-celled life in the water

Pressure: The continuous physical force applied on or to an object

Prism: A clear solid object with a triangular base and flat sides that breaks visible light into the colors of the spectrum.

Radiant energy: Energy traveling in the form of electromagnet waves, including radio waves, visible light, and x-rays

Ratio: A numerical relationship between two numbers in a specific proportion

Reservoir: A reserve of water in a lake or collecting pool stored for later use

Rotor: The part of a machine that rotates or spins

Saturation: The intensity of color

Seacow: The informal name for a manatee

Sirenian: Having to do with manatees and dugongs

Slash-and-burn: A way of adding more land for farming by cutting and burning the plants that grow there naturally; often done in forests

Solar panel: A flat structure lined with solar cells that absorb and collect the sun's energy which is then converted into electrical energy

Sphere: A round, solid figure with every surface point an equal distance from the midpoint

Statistics: The practice of collecting, organizing, and making sense of number data

Symmetry: A pattern in which exactly alike parts face opposite one another along a plane or center division or arranged equally around a midpoint

Traits: The qualities or characteristics of a living organism

Tuberculosis: A very contagious lung disease caused by bacteria

Turbine: A machine for generating power with a bladed wheel caused to rotate by water, wind, or gas passing over the blades

Vacuum: A sealed container holding no air or gases

Vapor: A gas formed after evaporation

Variable: The factor in an experiment that is controlled or changed

Vertex: The meeting point of two lines

Vibrate: To move or shake back and forth quickly

Voltage: The measure of the force of an electrical current

Voltmeter: An instrument for measuring the difference in potential between varying points of an electrical circuit in units called *volts*

Watershed: An area of land where water empties into rivers and lakes

POEM CREDITS

Joy Acey: "Capillary Action"; "Happiness in the Desert"; "Shots! Shots! Shots!"; copyright ©2014 by Joy Acey. Used with permission of the author.

Alma Flor Ada: "I Will Be a Chemist: Mario José Molina / Voy a ser químico: Mario Molina"; copyright ©2014 by Alma Flor Ada. Used with permission of the author.

Linda Ashman: "Life Story"; "No Hurry"; "Snake Traits"; copyright ©2014 by Linda Ashman. Used with permission of the author.

Jeannine Atkins: "Becoming Butterflies"; "Classroom in the Meadow"; "Elemental"; "Nursing Math"; "Playground Physics"; "Stone, Sea, and Silence"; "Uh Oh, Pluto"; copyright ©2014 by Jeannine Atkins. Used with permission of the author.

Carmen T. Bernier-Grand: "Computer Geek/Compu-nerdo"; "Mount St. Helens"; copyright ©2014 by Carmen T. Bernier-Grand. Used with permission of the author.

Robyn Hood Black: "Food for Thought"; "Printing, Pressed Beyond Words . . ."; "Rocky Rescue"; copyright ©2014 by Robyn Hood Black. Used with permission of the author.

Susan Blackaby: "Recycling"; "Resources Rule!"; "Solar Power"; "Scientific Inquiry"; copyright ©2014 by Susan Blackaby. Used with permission of the author.

Susan Taylor Brown: "A Dog's Hypothesis: Zoey's Guide to Getting More Goodies"; copyright ©2014 by Susan Taylor Brown. Used with permission of the author.

Joseph Bruchac: "What We Eat"; "Windfall in The Andrews Forest"; copyright ©2014 by Joseph Bruchac. Used with permission of the author.

Leslie Bulion: "Ocean Engine"; "Ocean Explorer Sylvia Earle"; "Testing My Hypothesis"; "Water Round"; copyright ©2014 by Leslie Bulion. Used with permission of the author.

Stephanie Calmenson: "Dog in a Storm"; "The Engineer"; copyright ©2014 by Stephanie Calmenson. Used with permission of the author.

F. Isabel Campoy: "Five O'Clock Rush" / "Prisas a las cinco"; "Power Blender and Aha!"/"Enchufa la batidora y ya esta!"; copyright ©2014 by F. Isabel Campoy. Used with permission of the author.

James Carter: "Hawking Time" copyright ©2014 by James Carter; "Now..." copyright ©2015 by James Carter. Used with permission of the author.

Kate Coombs: "Clouds"; "Glacier"; "Hands"; "Seeing School"; "Soil Inventory"; "Water"; copyright ©2014 by Kate Coombs. Used with permission of the author.

Cynthia Cotten: "Inquiry"; "Scientific Steps"; "What Is Science?"; copyright ©2014 by Cynthia Cotten. Used with permission of the author.

Kristy Dempsey: "Dinos in the Laboratory"; "Invention Intentions"; "(Super)Power: (to the) Point"; copyright ©2014 by Kristy Dempsey. Used with permission of the author.

Graham Denton: "Pass Me Those Ear Muffs"; copyright ©2014 by Graham Denton. Used with permission of the author.

Rebecca Kai Dotlich: "Water + Dirt ="; "The Crane Operator"; "What Is a Fossil?"; copyright ©2014 by Rebecca Kai Dotlich. Used with permission from Curtis Brown, Ltd.

Shirley Smith Duke: "Wondering Why"; "Citizen Scientist"; "At the Speed of Light"; "Teacher's Look"; "Patterns in Nature"; copyright ©2014 by Shirley Smith Duke. Used with permission of the author.

Margarita Engle: "Young & Old Together" / "Jovenes y viejos juntos"; "Discovery" / "Descubrimiento"; "Armor"; "We Need Green Seaweed!"; "Camouflage"; "A Biological Community" / "Una comunidad biologica"; "Tropical Rain Forest Sky Ponds"; "Shade-Grown"; "Pollination"/"Polinizacion"; copyright ©2014 by Margarita Engle. Used with permission of the author.

Douglas Florian: "Big Sun"; "Earth's Tilt"; copyright ©2014 by Douglas Florian. Used with permission of the author.

Betsy Franco: "Geologist"; "Moving for Five Minutes Straight"; "3-D"; copyright ©2014 by Betsy Franco. Used with permission of the author.

Carole Gerber: "Tornado!"; "Seeking Proof"; copyright ©2014 by Carole Gerber. Used with permission of the author.

Charles Ghigna: "Inherit Tense"; "Life Cycle"; copyright ©2014 by Charles Ghigna. Used with permission of the author.

Joan Bransfield Graham: "Superhero Scientist"; "Weather Map"; "Climate Versus Weather"; copyright ©2014 by Joan Bransfield Graham. Used with permission of the author.

Mary Lee Hahn: "Pumpkin Experiment"; "Dear Rachel Carson"; "The Lion and the House Cat"; "After I Made a Huge Mess with My Chemistry Set"; "Orion Nebula"; "Cancer"; copyright ©2014 by Mary Lee Hahn. Used with permission of the author.

Avis Harley: "Designing an Experiment: Will a Car Roll Faster Down a Steeper Slant?"; "Alexander Graham Bell"; "Lunar Eclipse"; copyright ©2014 by Avis Harley. Used with permission of the author.

David L. Harrison: "I Want to Know Why"; "My Robot"; copyright ©2014 by David L. Harrison. Used with permission of the author.

Terry Webb Harshman: "Queen of Night"; copyright ©2014 by Terry Webb Harshman. Used with permission of the author.

Juanita Havill: "Magic Show"; "Rings Not Letters"; "Space Yacht"; "The NEO Hunters"; copyright ©2014 by Juanita Havill. Used with permission of the author.

Esther Hershenhorn: "What Am I?"; copyright ©2014 by Esther Hershenhorn. Used with permission of the author.

Mary Ann Hoberman: "Trilobite"; copyright ©2014 by Mary Ann Hoberman. Used with permission of the author.

Sara Holbrook: "Water Engineered"; "Friction"; "Cool Food for Thought"; copyright ©2014 by Sara Holbrook. Used with permission of the author.

Patricia Hubbell: "Love Note to a Magnet"; "Roller Coaster Ride"; copyright ©2014 by Patricia Hubbell. Used with permission of the author.

Jacqueline Jules: "Protecting My Friend"; "Soda Machine Bite"; copyright ©2014 by Jacqueline Jules. Used with permission of the author.

Bobbi Katz: "The Rain Forest"; "Lunar Eclipse"; "Considering Copernicus"; copyright ©2014 by Bobbi Katz. Used with permission of the author.

X.J. Kennedy: "Metal Monster"; "Discovery"; copyright ©2014 by X.J. Kennedy. Used with permission of the author.

Julie Larios: "Albert Einstein"; "Did You Know?"; "My Experiment"; "Rachel Carson"; "Testing My Magnet"; copyright ©2014 by Julie Larios. Used with permission of the author.

Irene Latham: "Disaster Riddle in a Hurry"; "Disaster Riddle Under Pressure"; "Riddle for a Dry Day"; "Riddle for a Wet Day"; "Science Fair"; copyright ©2014 by Irene Latham. Used with permission of the author.

Renée M. LaTulippe: "Celsius Thermometer"; "Da Vinci Did It!"; "Driftwood Hut"; "Galileo Galilei"; "Lab Time"; "Let's All Be Scientists!"; "Pieces"; "Virtual Adventure"; copyright ©2014 by Renée M. LaTulippe. Used with permission of the author.

Debbie Levy: "Wiki Alert"; copyright ©2014 by Debbie Levy. Used with permission of the author.

J. Patrick Lewis: "The 'Black Leonardo'"; "What Should I Call It?"; copyright ©2014 by J. Patrick Lewis. Used with permission of Curtis Brown, Ltd.

George Ella Lyon: "Hand-Me-Downs"; "Meet Mr. Wizard"; copyright ©2014 by George Ella Lyon. Used with permission of the author.

Guadalupe Garcia McCall: "Cicada" / "Chicharra"; "Oh Water, My Friend" / "Ay agua, mi amiga"; "Sun-Kissed" / "Besado por el sol"; copyright ©2014 by Guadalupe Garcia McCall. Used with permission of the author.

Heidi Mordhorst: "Cicada Magic"; copyright ©2014 by Heidi Mordhorst. Used with permission of the author.

Marilyn Nelson: "A New Dinosaur"; copyright ©2014 by Marilyn Nelson. Used with permission of the author.

Kenn Nesbitt: "I Plan to Be an Astronaut"; "Lorenzo Liszt, Non-Scientist"; "My Project for the Science Fair"; copyright ©2014 by Kenn Nesbitt. Used with permission of the author.

Lesléa Newman: "First Science Project"; "The Leopard Cannot Change His Spots"; copyright ©2014 by Lesléa Newman. Used with permission of Curtis Brown, Ltd.

Eric Ode: "I Thought I Built a Dog House"; "Science Fair Day"; "Science Fair Project"; "When You Are a Scientist"; "Which Ones Will Float?"; copyright ©2014 by Eric Ode. Used with permission of the author.

Linda Sue Park: "Accidentally on Purpose"; "The Real Thing"; copyright ©2014 by Linda Sue Park. Used with permission of Curtis Brown, Ltd.

Ann Whitford Paul: "For the Science Fair..."; "Plates"; "Questions, Questions"; copyright ©2014 by Ann Whitford Paul. Used with permission of the author.

Greg Pincus: "Late Night Science Questions"; copyright ©2014 by Greg Pincus. Used with permission of the author.

Mary Quattlebaum: "Breakfast Alchemy"; copyright ©2014 by Mary Quattlebaum. Used with permission of the author.

Heidi Bee Roemer: "Going Bananas"; "Let Me Join You"; "Liquids Can't Contain Themselves"; "Questions That Matter"; "Thirsty Measures"; "Welcome to the Science Lab"; copyright ©2014 by Heidi Bee Roemer. Used with permission of the author.

Michael J. Rosen: "Titan in Man's Seaweed"; copyright ©2014 by Michael J. Rosen. Used with permission of the author.

Deborah Ruddell: "The Science Lab Pledge"; copyright ©2014 by Deborah Ruddell. Used with permission of Writers House LLC.

Laura Purdie Salas: "Can You Hear a Conch?"; "Go Fly a Kite"; "The Great Pyramid of Giza"; "Seashells in the Sky"; "The Shadow Grows (and Shrinks, and Grows)"; "Things to Do in Science Class"; "To the Eye"; "What Can You Make from Carbon?"; copyright ©2014 by Laura Purdie Salas. Used with permission of the author.

Michael Salinger: "Gears"; "Levers"; "No Penguins Here"; "Sound Waves"; "Squiggles"; copyright ©2014 by Michael Salinger. Used with permission of the author.

Glenn Schroeder: "Frisbee"; copyright ©2014 by Glenn Schroeder. Used with permission of the author.

Joyce Sidman: "Gravity"; copyright ©2014 by Joyce Sidman. Used with permission of the author.

Buffy Silverman: "Think of an Atom"; copyright ©2014 by Buffy Silverman. Used with permission of the author.

Marilyn Singer: "Lift"; "Photosynthesis"; copyright ©2014 by Marilyn Singer. Used with permission of the author.

Ken Slesarik: "Mrs. Sepuka's Classroom Pet"; "My Rock"; copyright ©2014 by Ken Slesarik. Used with permission of the author.

Eileen Spinelli: "Imagine Small"; "Thank You, Isaac Newton"; "What I Know about the Sun"; copyright ©2014 by Eileen Spinelli. Used with permission of the author.

Anastasia Suen: "I Have a Question"; "MMO"; "Rain Gauge"; copyright ©2014 by Anastasia Suen. Used with permission of the author.

Susan Marie Swanson: "Jane Goodall Begins a Speech"; "Story Rocks"; copyright ©2014 by Susan Marie Swanson. "Sound Waves at Breakfast"; ©2014, ©2018 by Susan Marie Swanson. Used with permission of the author.

Carmen Tafolla: "Everyday Astronaut" / "Un astronauta común"; "My WristRobot Pack"; copyright ©2014 by Carmen Tafolla. Used with permission of the author.

Holly Thompson: "Comet Hunter"; copyright ©2014 by Holly Thompson. Used with permission of the author.

Amy Ludwig VanDerwater: "How to Be a Scientist"; "Listen"; "Meter Stick"; "My Bean Plant"; "Prism"; "Sound Waves"; copyright ©2014 by Amy Ludwig VanDerwater. Used with permission of Curtis Brown, Ltd.

Lee Wardlaw: "Science Project"; copyright ©2014 by Lee Wardlaw. Used with permission of Curtis Brown, Ltd.

Charles Waters: "Foundation (Don't Rush It!)"; "Froggy"; "Mold"; copyright ©2014 by Charles Waters. Used with permission of the author.

April Halprin Wayland: "Can Our Eyes Fool Our Taste Buds?"; "I Like that Night Follows Day"; "Old Water"; "Zapped!"; copyright ©2014 by April Halprin Wayland. Used with permission of the author.

Carole Boston Weatherford: "Harbor Wave at Hilo: Tsunami Survivor"; copyright ©2014 by Carole Boston Weatherford. Used with permission of the author.

Steven Withrow: "What Makes a Turbine Turn"; "Moving to Atlantis City, 2112"; copyright ©2014 by Steven Withrow. Used with permission of the author.

Allan Wolf: "Step Outside. What Do You See?"; copyright ©2014 by Allan Wolf. Used with permission of the author.

Virginia Euwer Wolff: "That Dish Thing"; copyright ©2014 by Virginia Euwer Wolff. Used with permission of Curtis Brown, Ltd.

Janet Wong: "Auntie V's Hybrid Car"; "Backwards"; "The Brink"; "Changes"; "The Class Plant"; "Computer Models"; "Crazy Data Day"; "Dr. Lee"; "Fossil Fuels"; "Frames Per Second (fps)"; "Future Dreams Idea #63"; "Game Programmer"; "Grafting"; "Hello, Hello!"; "Hurricane Hideout"; "Looking at the Sky Tonight"; "Microwave Oven"; "My Photo Experiment"; "Our Truck"; "Paper Airplanes"; "Push Power"; "Shen Kuo"; "Sink or Float"; "Stopwatch"; "Sugar Water"; "Take Backs"; "This Week's Weather"; "Tinker Time"; copyright ©2014 by Janet S. Wong. Used with permission of the author.

Jane Yolen: "Alligator with Fish"; "The Lament of Lonesome George"; "Tide Pool"; "What Is a Foot?"; "Butterfly Garden"; "Seven Words about an Alligator"; "Snack"; copyright ©2014 by Jane Yolen. Used with permission of Curtis Brown, Ltd.

ACKNOWLEDGMENTS

Poets observe nature, explore the physical world, and ask questions about the universe—just like scientists! So compiling an anthology of poems that are science-themed was a compelling invitation for all of the poets who participated in this publication, and we are so grateful for their wonderful, diverse contributions. Without the poets, we have nothing. Please get to know them and their books—and look for their individual websites.

It was very important to "get the science right," so we also took great care to "vet" the poems for science content. Here's a gigantic *thank you* to our several science experts who reviewed our poems (and the Take 5! activities that you can find for these poems in the Teacher Edition of *The Poetry Friday Anthology for Science*). We especially want to recognize Shirley Smith Duke, Britt Bothe, Kathleen Hoke, and Mark VanDerwater for their comments to an early version of the companion Teacher Edition to this book, as well as Seymour Simon for his early praise and continuing support of our efforts to promote poetry and science together. We also appreciate the close reading, additional research, and editorial support provided by the amazing Shirley Smith Duke (it bears repeating!), Cynthia Alaniz, and Emily Vardell. Finally, we want to thank artists Frank Ramspott and Bug Wang for their engaging illustrations that add so much to these poems.

We hope that this book makes you wonder about the natural world and also treasure the power of words and language. If you liked this book, please help spread the word!

With sincerest appreciation,

Sylvia and Janet

PomeloBooks.com

ABOUT THE AUTHORS

Sylvia M. Vardell is Professor at Texas Woman's University and teaches courses in children's and young adult literature. She has published five books on literature for children, as well as over 25 book chapters and 100 journal articles. Her current work focuses on poetry for children, including a regular blog, PoetryforChildren, since 2006. Her favorite part of science is geology and she loves collecting rocks, minerals, and shells.

Janet S. Wong is a graduate of Yale Law School and a former lawyer who became a children's poet. Her work has been featured on *The Oprah Winfrey Show* and other shows; she is the author of 30 books for children and teens on chess, creative recycling, yoga, superstitions, driving, and more. She loves watching short science and engineering videos (on topics like making a hoverboard).

K-5 POETRY
(TEACHER EDITION)

218 poems by 76 poets about school, pets, favorite sports, weather, food, friends, free time, vacation, books, and more, with *Take 5!* mini-lessons for each poem

MIDDLE SCHOOL POETRY
(TEACHER EDITION)

110 poems by 71 poets about new schools, coping with family, playing soccer, and texting friends, with *Take 5!* mini-lessons *(an NCTE Poetry Notable Book)*

K-5 SCIENCE POETRY
(TEACHER EDITION)

A companion to this book, with 218 science poems and *Take 5!* mini-lessons that teachers, librarians, and parents can use to connect each poem to the NGSS or state standards *(an NSTA Recommends book)*

HOLIDAY POETRY EDITION
(TEACHER/LIBRARIAN & CHILDREN'S EDITIONS)

156 poems by 115 poets in English + Spanish about celebrations from Halloween and New Year's Eve to National Pizza Week and National Bike Month

For more information about *The Poetry Friday Anthology*® series, including the Teacher Edition of *The Poetry Friday Anthology for Science*, please visit **PomeloBooks.com**.

Made in the USA
Middletown, DE
26 April 2019